Numerical Relativity: Starting from Scratch

Numerical relativity has emerged as the key tool to model gravitational waves – recently detected for the first time – emitted when black holes or neutron stars collide. This book provides a pedagogical, accessible, and concise introduction to the subject. Relying heavily on analogies with Newtonian gravity, scalar fields, and electromagnetic fields, it introduces key concepts of numerical relativity in contexts familiar to readers without prior expertise in general relativity. Readers can explore these concepts by working through numerous exercises and can see them "in action" by experimenting with the accompanying Python sample codes and so develop familiarity with many techniques commonly employed by publicly available numerical relativity codes. This is an attractive, student-friendly resource for short courses on numerical relativity and it also provides supplementary reading for courses on general relativity and computational physics.

THOMAS W. BAUMGARTE is the William R. Kenan Jr. Professor of Physics at Bowdoin College in Brunswick, Maine. His work in numerical relativity and relativistic astrophysics has been recognized with prizes and fellowships from the Guggenheim Foundation, the Humboldt Foundation, the American Physical Society, and the Simons Foundation. Stuart Shapiro and Thomas Baumgarte previously co-authored the graduate-level text *Numerical Relativity: Solving Einstein's Equations on the Computer* (Cambridge University Press, 2010).

STUART L. SHAPIRO is a Professor of Physics and Astronomy at the University of Illinois at Urbana-Champaign. He is a leading scientist in theoretical astrophysics and general relativity and has been awarded numerous prizes and honors for his research and teaching, including Sloan and Guggenheim fellowships, IBM Supercomputing awards, and the Hans A. Bethe Prize of the American Physical Society, to which he was elected Fellow. Stuart Shapiro has published over 400 research papers and, in addition to his writing with Thomas Baumgarte, is co-author of the classic text *Black Holes, White Dwarfs and Neutron Stars: The Physics of Compact Objects* (Wiley, 1983).

Numerical Relativity: Starting from Scratch

Thomas W. Baumgarte
Bowdoin College

Stuart L. Shapiro
University of Illinois at Urbana-Champaign

CAMBRIDGE
UNIVERSITY PRESS

CAMBRIDGE
UNIVERSITY PRESS

University Printing House, Cambridge CB2 8BS, United Kingdom

One Liberty Plaza, 20th Floor, New York, NY 10006, USA

477 Williamstown Road, Port Melbourne, VIC 3207, Australia

314–321, 3rd Floor, Plot 3, Splendor Forum, Jasola District Centre, New Delhi – 110025, India

79 Anson Road, #06–04/06, Singapore 079906

Cambridge University Press is part of the University of Cambridge.

It furthers the University's mission by disseminating knowledge in the pursuit of education, learning, and research at the highest international levels of excellence.

www.cambridge.org
Information on this title: www.cambridge.org/9781108844116
DOI: 10.1017/9781108933445

© Thomas W. Baumgarte and Stuart L. Shapiro 2021

First published 2021

Printed in Singapore by Markono Print Media Pte Ltd

A catalogue record for this publication is available from the British Library.

Library of Congress Cataloging-in-Publication Data

Names: Baumgarte, Thomas W., author. | Shapiro, Stuart L. (Stuart Louis), 1947– author.
Title: Numerical relativity : starting from scratch / Thomas W. Baumgarte, Stuart L. Shapiro.
Description: [Updated edition]. | New York : Cambridge University Press, 2021. | Includes bibliographical references and index.
Identifiers: LCCN 2020040206 (print) | LCCN 2020040207 (ebook) | ISBN 9781108844116 (hardback) | ISBN 9781108928250 (paperback) | ISBN 9781108933445 (epub)
Subjects: LCSH: General relativity (Physics) | Einstein field equations. | Numerical calculations.
Classification: LCC QC173.6 .B38 2021 (print) | LCC QC173.6 (ebook) | DDC 530.11–dc23
LC record available at https://lccn.loc.gov/2020040206
LC ebook record available at https://lccn.loc.gov/2020040207

ISBN 978-1-108-84411-6 Hardback
ISBN 978-1-108-92825-0 Paperback

Additional resources for this publication at www.cambridge.org/NRStartingfromScratch

Contents

Preface

General relativity, Einstein's theory of relativistic gravitation, provides the foundation upon which black holes and neutron stars, compact binary mergers, gravitational waves, cosmology, and all other strong-field gravitational phenomena are constructed. Solutions to Einstein's equations, except for a few special cases characterized by high degrees of symmetry, cannot be achieved by analytical means alone for many important dynamical scenarios thought to occur in nature. With the aid of computers, however, it is possible to tackle these highly nonlinear equations numerically in order to examine these scenarios in detail. That is the main objective of *numerical relativity*, the art and science of casting Einstein's equations in a form suitable for applying general relativity to physically realistic, high-velocity, strong-field dynamical systems and then designing numerical algorithms to solve them on a computer.

Until a little over a dozen years ago, there did not exist a single textbook focusing on numerical relativity. Since then, however, several volumes have appeared, including those by Bona and Palenzuela (2005); Alcubierre (2008); Bona et al. (2009); Gourgoulhon (2012) and Shibata (2016), as well as our own textbook, Baumgarte and Shapiro (2010). One might wonder whether there is any need to add yet another book to this list. However, we believe that this short volume complements the above references rather than duplicating any one of them, as we will explain below.

Several events and developments have occurred since the appearance of most of the above treatises. Most significantly, gravitational waves have been detected directly for the first time, confirming a major prediction of Einstein's theory and launching the field of gravitational wave astronomy. In 2015, a gravitational wave signal originating from a binary black-hole system that coalesced in a distant galaxy about a billion years ago was observed on Earth by the Laser Interferometer Gravitational Wave Observatory (LIGO). This detection represented a true milestone in gravitational physics, which was acknowledged by the awarding of a Nobel Prize in 2017 to Barry Barish, Kip Thorne, and Rainer Weiss, the physicists who led the construction of LIGO and the search for gravitational waves. Since then, additional gravitational wave signals from binary black-hole mergers have been observed and analyzed. In 2017 the first gravitational wave signal from the merger of a binary neutron star was detected and it was accompanied by the observation of a gamma-ray burst, together with other electromagnetic radiation. A gravitational wave detection simultaneous with the observation of counterpart electromagnetic radiation represents the holy grail of "multimessenger astronomy", a golden moment in this growing field. We will discuss some of these observations in more detail in Sections 1.3.2 and 5.1.

These ground-breaking discoveries were met with profound interest and fascination by scientists and the general public alike. They also helped to attract a broader community to the field of numerical relativity, which has emerged as the major tool needed to predict and interpret gravitational wave signals from cosmic events, such as compact-binary mergers. In the past, numerical relativity was practiced predominantly by scientists with appreciable expertise in both general relativity and computational physics, but these new discoveries have prompted broader segments of the physics and astrophysics communities to make use of this tool.

In an independent development, rather sophisticated and mature numerical relativity codes have become publicly and freely available, thereby enabling broader access to the tools of numerical relativity. While, in the past, most numerical relativity groups developed and utilized their own codes, many individuals both within and outside traditional numerical relativity circles are beginning to rely on these "open-source" community codes for their investigations.

This broadening of interest in numerical relativity coupled with the greater availability of numerical relativity codes has spurred a growth in the audience for numerical relativity textbooks. Traditionally, the readers of such books, already well-trained in basic general relativity, are interested in a comprehensive and thorough development of the subject, including a complete treatment of the mathematical underpinnings and full derivation of all the equations. The textbooks listed above are aimed at that readership. However, many researchers newly attracted to numerical relativity, including current and future users of community codes, may not have that background and may find the currently available books on numerical relativity inaccessible. We suspect, nevertheless, that many such users may want to gain at least an intuitive understanding of some of the variables, key equations, and basic algorithms often encountered in numerical relativity, rather than using community codes solely as "black boxes". This book is aimed at those readers.

The purpose of this book, then, is to provide an exposition to numerical relativity that is accessible, does not rely on an intimate prior knowledge of general relativity, and instead builds on familiar physics and intuitive arguments. Our hope is that it will help users of numerical relativity codes with limited expertise in general relativity "look under the hood", allowing them to take a glance at what lies inside the codes and algorithms and to gain a level of familiarity with numerical relativity, so that they can use these codes "on the road" with greater confidence, understanding, and expectation of reliability.

One of us (TWB) has taught at several summer schools on numerical relativity and relativistic astrophysics over the last few years, and this experience has certainly helped to motivate our approach. The students, many of whom had limited experience with general relativity, were best served by skipping over some derivations and technical details and spending more time on motivating the physical and geometrical meaning of the variables and equations encountered in a 3+1 decomposition of Einstein's field equations. This decomposition involves splitting four-dimensional spacetime into three-dimensional space plus one-dimensional time and recasting the equations accordingly. A brief introduction to general relativity is required first, of course, and we provide one here, together with an introduction to tensors and elements of differential geometry. However, since many concepts encountered in the 3+1 decomposition of the equations can be introduced in the much more familiar context of electrodynamics, it is possible to disentangle many issues associated with this 3+1 splitting from the special subtleties associated with general relativity, prior to discussing the latter. This is the approach we adopt in this volume.

Our goal, then, is neither to provide a comprehensive review of the entire field nor to present every derivation in complete detail; we essentially did just that in Baumgarte and Shapiro (2010). Instead, here we hope to provide a more accessible introduction to some concepts and quantities encountered in numerical relativity, leaning heavily on analogies with Newtonian mechanics, scalar fields, and electrodynamics. The hope is that readers with a solid background in basic mechanics, electrodynamics, special relativity, and vector calculus, but with no familiarity, or only limited familiarity, with general relativity and differential geometry, will be able to follow the development. They should be able to gain an intuitive understanding of the concepts and variables arising in numerical relativity codes, recognize the structure of the equations encountered in a 3+1 decomposition, and appreciate some techniques often employed in their solution. Ideally, such a reader will then be motivated and able to consult any book in the list above for a more detailed and comprehensive discussion of one or more topics.

One of us (SLS) used an early draft of this book to teach a seminar on numerical relativity to a small group of junior and senior physics majors with little formal background in general relativity. They all seemed to warm to the approach and were able to absorb the basic ideas, as demonstrated by our weekly discussions and by their solving many of the exercises to be found throughout the volume.

Another feature that distinguishes this book from other textbooks on numerical relativity is that we include actual numerical codes. The broad availability of Python and its many libraries makes it easier than in the past to provide small sample scripts that, despite being limited in scope, highlight some features of many numerical relativity codes. We provide two such codes that illustrate the two basic types of problems encountered in numerical relativity: initial-value and evolution problems. These codes allow the reader to watch and explore some common computational algorithms "at work".

This volume is organized as follows. In Chapter 1 we sketch Einstein's theory of general relativity, relying extensively on comparisons with Newtonian gravity. Clearly, a single chapter on general relativity cannot possibly replace a full course or book on the subject, but we nevertheless hope that it will provide a quick review for readers with some knowledge of general relativity and, more importantly, simultaneously serve as a brief introduction and overview for readers unfamiliar with the subject.

In Chapter 2 we discuss how Einstein's field equations in standard form can be recast to treat an initial-value problem in general relativity that can then be solved numerically. This step employs the 3+1 decomposition of spacetime that we mentioned above. We draw heavily on analogies with both scalar fields and electrodynamics, using these more familiar settings to introduce many of the geometrical objects encountered in numerical relativity. The result is a split of the field equations into constraint and evolution equations.

In Chapter 3 we introduce some strategies for solving the constraint equations, including conformal decompositions. Solutions to the constraint equations describe the fields at one instant of time and serve as initial data for dynamical evolutions.

In Chapter 4 we then discuss evolution calculations. This entails finding formulations of the evolution equations and coordinate conditions that result in stable numerical behavior.

In Chapter 5 we review simulations of binary black-hole mergers in vacuum as an example of such calculations. Simulations like those summarized here have played a crucial role in the detection and analysis of the gravitational waves mentioned above.

This book also includes several appendices. Appendix A summarizes some basic tensor concepts, including the covariant derivative.

Appendix B introduces some computational techniques commonly used in numerical relativity. In particular, this appendix includes the two Python scripts to which we referred above; one solves the (elliptic) constraint equations of Chapter 3 for an isolated black hole, and the other solves the (hyperbolic) evolution equations of Chapter 4 for a propagating electromagnetic wave. Both programs can also be downloaded from www.cambridge.org/NRStartingfromScratch. Here readers can vary the input parameters, run and modify the codes, and evaluate the output using the diagnostic graphing and animation routines that are also provided.

In the interest of providing a coherent development while keeping this volume short, we focus chiefly on vacuum spacetimes, i.e. those containing black holes and/or gravitational waves but no matter or other nongravitational fields as source terms in the Einstein equations. Stated differently, we concentrate on the left-hand side of Einstein's equations (the "geometry"), setting the right-hand side (the "matter and nongravitational fields") to zero. However, in Appendix C we extend our treatments of scalar and electromagnetic fields to identify their stress–energy tensors, which can serve as sources for Einstein's equations, and show how general relativity imposes the familiar equations of motion for these nongravitational fields.

Appendix D provides a short summary of some of the most important mathematical results in differential geometry, general relativity, and the 3+1 decomposition that are encountered in numerical relativity.

For pedagogical purposes we have inserted a total of 74 exercises scattered throughout this book, some of which are numerical. They are designed to help readers test and develop their basic understanding of the material. In some cases the exercises build on each other, so in Appendix E we provide answers to those that are needed to tackle subsequent exercises.

We are deeply indebted to more people than we could possibly list here, including all those colleagues, teachers, and students from whom, over the years and decades, we have learned almost everything we know. Some of these have also had a direct impact on this book: We would like to thank Parameswaran Ajith, Andreas Bauswein, Stephan Rosswog, and Bjoern Malte Schaefer for inviting TWB to lecture at their summer schools, as well as the students at those schools for their feedback. We would also like to thank Elizabeth Bennewitz, Béatrice Bonga, Kenneth Dennison, Steven Naculich, and Constance Shapiro, who read sections of this book and provided helpful comments, and Eric Chown and Zachariah Etienne for their help with the Python scripts. Several reviewers gave us valuable feedback; we would like to thank Ulrich Sperhake in particular for his detailed report and helpful suggestions. We would like to express our gratitude to the seminar participants at the University of Illinois – Michael Mudd, Kyle Nelli, Minh Nguyen, and Samuel Qunell – who provided invaluable feedback on an early draft of this volume, as well as solutions to most of the exercises. We also gratefully acknowledge the National Science Foundation (NSF), the National Aeronautics and Space Administration (NASA), and the Simons Foundation for funding our research. Many figures in this volume were produced using Mayavi software. Our acknowledgements would not be complete, however, without thanking our wives and families for their enduring and loving support.

Finally, we hope we will be forgiven for referring quite often to our previous textbook throughout this volume for further details on various topics. As we are most familiar with that book, it is expedient for us to do so and we hope that other authors will not feel in any way slighted by our taking advantage of our previous exposition.

Thomas W. Baumgarte
Stuart L. Shapiro

1

Newton's and Einstein's Gravity

It is impossible to introduce general relativity in just one chapter, and we will not attempt that here.[1] *Instead, we will review some properties of Newtonian gravity in this chapter, and will then develop some key ideas, concepts, and objects of general relativity by retracing the very same steps that we follow in our review of the Newtonian theory.*

1.1 A Brief Review of Newton's Gravity

According to Sir Isaac Newton, gravity makes objects follow a curved trajectory. The origin of this curvature is a gravitational force $\mathbf{F}^N$ that acts on objects with non-zero mass. The gravitational force acting on an object is proportional to the mass m_G of this object; we may therefore write the force as

$$\mathbf{F}^N = m_G \mathbf{g}, \tag{1.1}$$

where $\mathbf{g}$ is the gravitational field created by all other objects. Note that we have decorated the gravitational mass with a subscript G in order to distinguish it from the inertial mass m_I, which we will introduce in equation (1.4) below. From equation (1.1) we see that the gravitational mass m_G describes how strongly an object couples to the gravitational field created by other masses. In electrodynamics, the equation equivalent to (1.1) would be $\mathbf{F}^N = q\mathbf{E}$, and so we see that the electrodynamic cousin of the gravitational mass is the charge q.

---

[1] See, e.g., Misner et al. (2017); Weinberg (1972); Schutz (2009); Wald (1984); Hartle (2003); Carroll (2004); Moore (2013) for some excellent textbooks on general relativity.

The gravitational field **g** is always irrotational, meaning that its curl vanishes. We refer to such fields as "conservative" and recall that they can always be written as the gradient of a scalar function. For the gravitational field we introduce the *Newtonian potential* Φ and write

$$\mathbf{g} = -\mathbf{D}\Phi. \qquad (1.2)$$

Here the negative sign follows convention, and we use the symbol **D** for the three-dimensional, spatial gradient operator because we will reserve the nabla operator ∇ for four-dimensional gradients. We will consider Φ the "fundamental quantity" in Newtonian theory, the quantity that encodes gravitational interactions.

Throughout this text we will often use index notation. Using an index i to denote the components in the vectors in equation (1.2), for example, would result in

$$g_i = -D_i\Phi = -\frac{\partial}{\partial x^i}\Phi = -\partial_i\Phi, \qquad (1.3)$$

where the components D_i of the gradient **D** are just spatial partial derivatives. We also introduce the shorthand notation ∂_i to denote the partial derivative with respect to the spatial coordinate x^i. Here the index i runs from 1 to 3, representing the three spatial directions. In Cartesian coordinates, for example, x would be represented by $i = 1$, y by $i = 2$, and z by $i = 3$. We could also ask whether it matters that the indices are "downstairs" rather than "upstairs" – it does, see Appendix A! – but we will ignore that subtlety for a little while.

Newton's second law tells us that objects accelerate in response to a gravitational force $\mathbf{F}^N$ according to the Newtonian equations of motion,

$$m_I\mathbf{a} = \mathbf{F}^N = -m_G\mathbf{D}\Phi \quad \text{or} \quad m_I a_i = F_i^N = -m_G D_i\Phi. \qquad (1.4)$$

Here we have decorated the mass m_I on the left-hand side with the subscript I in order to emphasize that this is the inertial mass: it describes how strongly an object resists being accelerated. The gravitational mass m_G and the inertial mass m_I therefore describe two completely independent internal properties of objects, and *a priori* it is not at all clear why the two should be the same. Note that we can define the gravitational mass m_G without any involvement of acceleration; likewise, the inertial mass m_I measures an object's response to any force, not just a gravitational force. In fact, in the electrodynamic analogue the two related quantities m_I and q do not even have the same units. And yet Newton assures us that the two masses are indeed equal,

$$m_G = m_I,$$
(1.5)

so that (1.4) reduces to

$$a_i = \frac{dv_i}{dt} = -D_i \Phi,$$
(1.6)

where v_i is the object's spatial velocity. Equation (1.6) is in accordance with Galileo Galilei's remarkable observation that all objects in a gravitational field Φ fall at the same rate, independently of their mass. We refer to this observation, which will motivate our development of general relativity in Section 1.2 below, as the (weak) equivalence principle.

It seems intuitive that we ought to be able to measure gravitational fields from the acceleration of falling objects but, as an immediate consequence of the equivalence principle, this is not necessarily the case. Consider, for example, performing experiments – however short-lived they might be – in a freely-falling elevator. In particular, we could let an object drop, say a marshmallow. Since the elevator and the marshmallow drop at the same rate,[2] we would not observe the marshmallow moving at all. Newton accounts for this by introducing "fictitious" forces and corresponding "fictitious" accelerations; in the frame of the elevator, for example, we should replace (1.6) with

$$a_i = -\partial_i \Phi - a_i^{\text{elevator}} \quad \text{(in falling elevator)}.$$
(1.7)

Here both $\partial_i \Phi$ and the elevator's acceleration a_i^{elevator} are measured in what Newton referred to as an inertial frame – namely a frame in which all fictitious forces vanish and free particles move at constant velocity. Since the two terms $\partial_i \Phi$ and a_i^{elevator} cancel each other in equation (1.7), we have $a_i = 0$ in the falling elevator. Measuring the spatial acceleration of a freely-falling object is evidently not a frame-independent measure of the gravitational field.

Does that mean that there is no frame-independent measure of gravitational fields? Is there any way to measure gravitational fields in the freely-falling elevator, for example? It turns out that there is, as long as the gravitational fields are inhomogeneous, i.e. not constant and not independent of position. To see this, consider dropping not one but two marshmallows, as shown in Fig. 1.1. If the two marshmallows

---

[2] In the limit that both are small compared with the scale of spatial variations of the gravitational field.

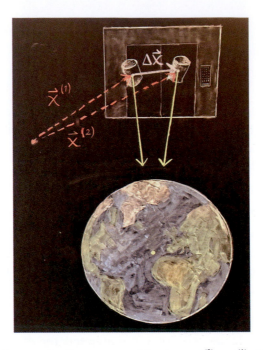

Figure 1.1 Two marshmallows, separated by $\Delta\mathbf{x} = \mathbf{x}^{(2)} - \mathbf{x}^{(1)}$, falling inside a freely-falling elevator. Both marshmallows fall towards the center of the Earth. Local variations in the gravitational forces, which we refer to as tidal forces, result in changes in the deviation $\Delta\mathbf{x}$. (Drawing by Julia Baumgarte.)

start out at the same horizontal level, their separation will decrease in time as they both drop towards the center of the Earth. If, on the other hand, one starts out above the other, the upper one will feel a slightly smaller gravitational attraction to the Earth than the lower one, and their separation will therefore increase. The Earth's *tidal forces* result in a changing separation of the two marshmallows, which we can measure. As we will see, these tidal forces are related to the second derivatives of the potential Φ.

Say that the position vector of one marshmallow has components $x_i^{(1)}(t)$ while the other has components $x_i^{(2)}(t) = x_i^{(1)}(t) + \Delta x_i(t)$, where Δx_i measures the time-dependent deviation between the two marshmallows. The key idea is that the deviation Δx_i satisfies an equation that is independent of the reference frame. We can evaluate

$$\frac{d^2\Delta x_i}{dt^2} = \frac{d^2}{dt^2}(x_i^{(2)} - x_i^{(1)}) = a_i^{(2)} - a_i^{(1)} = -(D_i\Phi)^{(2)} + (D_i\Phi)^{(1)}$$
(1.8)

in either frame, i.e. either from (1.6) or (1.7), and obtain the same result since the additional term a_i^{elevator} in (1.7) is the same for both marshmallows and hence drops out. The term $D_i\Phi$, however, has to be evaluated at $x_i^{(1)}$ for one marshmallow and at $x_i^{(2)}$ for the other. Let's assume that the two marshmallows are close to each other, so that the Δx_i are small (in comparison to the length scales over which Φ changes) and we may use a leading-order Taylor expansion in Cartesian coordinates to express $(D_i\Phi)^{(2)} = (\partial_i\Phi)^{(2)}$ as

$$(\partial_i\Phi)^{(2)} = (\partial_i\Phi)^{(1)} + \sum_{j=1}^{3}\Delta x^j(\partial_j\partial_i\Phi)^{(1)} + \mathcal{O}(\Delta x^2)$$

$$= (\partial_i\Phi)^{(1)} + \Delta x^j(\partial_j\partial_i\Phi)^{(1)} + \mathcal{O}(\Delta x^2).$$
(1.9)

In the middle expression we have written the sum over the indices j explicitly; in the last expression we have introduced the "Einstein summation convention", by which we automatically sum over all allowed values of any two repeated indices, one upstairs and one downstairs. Expressed in index notation, the dot product between two vectors $\mathbf{v}$ and $\mathbf{w}$, for example, becomes $\mathbf{v} \cdot \mathbf{w} = v^i w_i$. We similarly have $x_i x^i = x^2$, and $\partial_j(x_i x^i) = 2x_j$. Inserting (1.9) into (1.8) and ignoring higher-order terms we obtain

$$\frac{d^2\Delta x_i}{dt^2} = -\Delta x^j(\partial_j\partial_i\Phi),$$
(1.10)

where we have now dropped the superscript 1. This is known as the *Newtonian tidal deviation equation*. We conclude that, unlike the acceleration of a single marshmallow, which is computed from the first derivatives $\partial_i\Phi$, the relative acceleration between the two marshmallows, computed from the second derivatives $\partial_j\partial_i\Phi$, is independent of the reference frame. The former describe the gravitational field, while the latter describe derivatives of the gravitational field, which are related to tidal forces – in fact, we will refer to the object

$$\mathcal{R}_{ij} \equiv \partial_i\partial_j\Phi$$
(1.11)

as the *Newtonian tidal tensor*. The word "tensor" here is really short-hand for "rank-2 tensor". We define tensors properly in Appendix A; for our purposes here it is sufficient to say that a rank-n tensor carries n indices. Special cases are scalars, which are rank-0 tensors and have no indices, vectors, which are rank-1 tensors with one index, and the rank-2 tensors that we have just met. We can display a rank-2 tensor as a matrix, where the rows correspond to one index and the columns to the other. With the help of the tidal tensor (1.11), the Newtonian deviation equation (1.10) takes the compact form

$$\frac{d^2 \Delta x_i}{dt^2} = -\mathcal{R}_{ij} \Delta x^j. \tag{1.12}$$

We have not yet discussed the equation that governs the gravitational potential Φ itself. From Newton's universal law of gravitation, we can show that Φ satisfies the Poisson equation

$$D^2 \Phi = 4\pi G \rho_0, \tag{1.13}$$

where ρ_0 is the rest-mass density,[3] D^2 is the Laplace operator, and G is the gravitational constant, which we will set to unity shortly. The Poisson equation determines the fundamental quantity in Newtonian gravity, which is in fact a field – we should therefore think of the Poisson equation as the *Newtonian field equation*.

We now notice an interesting connection between the Newtonian field equation and the tidal tensor. Writing out the left-hand side of (1.13) in Cartesian coordinates we find

$$D^2 \Phi = \frac{\partial^2 \Phi}{\partial x^2} + \frac{\partial^2 \Phi}{\partial y^2} + \frac{\partial^2 \Phi}{\partial z^2} = \mathcal{R}^i{}_i, \tag{1.14}$$

where we invoke the Einstein summation on the right-hand side. We are, admittedly, being sloppy with regard to whether indices are upstairs or downstairs and will soon see how to clean this up, but in Cartesian coordinates the above is nevertheless rigorous. The important result is that we can write the Laplace operator acting on Φ as the so-called *trace* of the tidal tensor $\mathcal{R} \equiv \mathcal{R}^i{}_i$, i.e. the sum over its diagonal components (those for which both indices are the same). In terms of this trace we can then write the Newtonian field equation as

$$\mathcal{R} = 4\pi G \rho_0. \tag{1.15}$$

---

[3] We use the subscript 0 to distinguish the *rest-mass* density ρ_0 from the *total* mass–energy density ρ, which we will encounter in the context of general relativity. The latter contains contributions from the rest-mass energy *and* the bulk kinetic plus internal energy.

Exercise 1.1 Consider a point mass M located at a position $\mathbf{r}_M$. At a position $\mathbf{r}$, the Newtonian potential Φ created by the point mass is

$$\Phi(\mathbf{r}) = -\frac{GM}{s}, \tag{1.16}$$

where we have defined $\mathbf{s} = \mathbf{r} - \mathbf{r}_M$ and $s = |\mathbf{s}|$.

(a) Show that the tidal tensor can be written in the compact form

$$\mathcal{R}_{ij} = \frac{GM}{s^5}\left(\delta_{ij}s^2 - 3s_i s_j\right), \tag{1.17}$$

where δ_{ij} is the Kronecker delta: it takes the value one when both indices are equal and zero otherwise.

(b) Choose a (Cartesian) coordinate system in which $\mathbf{r}_M = (0, 0, z_M)$ and find all non-zero components of the tidal tensor at the origin, $\mathbf{r} = (0, 0, 0)$.

(c) Consider two particles close to the origin that are separated by a distance $\Delta z \ll z_M$ along the direction vector towards M, i.e. $\Delta x^i = (0, 0, \Delta z)$, and find $d^2\Delta z/dt^2$. Then consider two particles close to the origin that are separated by a distance $\Delta x \ll z_M$ in a direction orthogonal to the direction towards M, i.e. $\Delta x^i = (\Delta x, 0, 0)$, and find $d^2\Delta x/dt^2$. If all goes well, this should explain why objects in the gravitational field of a companion, e.g. the Earth in the gravitational field of the Moon, are elongated in the direction of the companion and compressed in the orthogonal directions.

(d) Show that the trace $\mathcal{R} = \mathcal{R}^i{}_i$ of the tidal tensor (1.17) vanishes for all $s > 0$, as expected from the field equation (1.15).

This brings us to the following outline of our quick tour of Newtonian gravity:

1. The fundamental quantity in Newtonian gravity is the Newtonian potential Φ.
2. Trajectories of objects are, by virtue of the equations of motion (1.6), governed by the first derivatives of the fundamental quantity.
3. Deviations between nearby trajectories are governed by the second derivatives of the fundamental quantity, i.e. the tidal tensor, see (1.12).
4. The field equation (1.15) relates the trace of the tidal tensor to matter densities, i.e. the source of the gravitational field.

As it turns out, we can take a brief tour of Einstein's gravity by retracing these exact steps – in fact, every one of the items in the above list will have a corresponding subsection in our introduction to general relativity. As both a preview and summary we list all important gravitational quantities, and relate the Newtonian terms to their relativistic analogues, in Box 1.1. Note in particular that, in Einstein's theory, each object or

---

Box 1.1 Important gravitational quantities

	Relation to fundamental quantity	Newton	Einstein
Fundamental quantity	—	potential Φ	metric g_{ab}
Equation of motion	first derivative	$D_i \Phi$	Christoffel symbols $^{(4)}\Gamma^a_{\ bc}$
Geodesic deviation	second derivatives	$D_i D_j \Phi$	Riemann tensor $^{(4)}R^a_{\ bcd}$
Field equation l.h.s.	trace of 2nd derivs.	$D^2 \Phi$	Einstein tensor G_{ab}
Field equation	—	$D^2 \Phi = 4\pi G \rho_0$	$G_{ab} = \dfrac{8\pi G}{c^4} T_{ab}$

---

equation is a tensor whose rank is two higher than its counterpart in Newton's theory.

1.2 A First Acquaintance with Einstein's Gravity

Recall from our discussion above that, in order to reproduce Galilei's observation that all objects fall at the same rate, we had to assume the equivalence (1.5) of the gravitational and inertial masses m_G and m_I. Granted, anything else would probably seem very odd to most readers, but that is only because we are not used to even considering the possibility that they might be different – in fact, we rarely even use different symbols for the two intrinsic properties. As physicists, we "grew up" with the notion that they are the same! And yet, as we discussed above, the gravitational mass and the inertial mass measure two completely independent intrinsic properties, and there is no *a priori* reason at all for them to be the same. Newtonian gravity is therefore based on a remarkable coincidence: in order to reproduce a global observation, namely that the trajectory of an object is independent of the object's mass, all objects must share an individual property, namely that their two masses are equal.

We can either accept this as a coincidence or, as Albert Einstein did, we can think of this as a "smoking gun", suggesting that there is a deeper reason for this property of gravity. Einstein realized that, instead of relying on individual properties of individual objects, it seems much more natural to assume that the trajectory is not caused by the properties of the objects at all but rather by something that is independent of the objects, something that two nearby objects share.

Consider dropping, next to each other, a marshmallow and a bowling ball, two objects with very different masses that nevertheless will fall at the same rate.[4] What these two objects share is the space through which they fall. A property of this space that could affect the trajectory of the two objects is its *curvature*. This is indeed the basic principle of Einstein's general relativity: Einstein assumed that space is curved, with the curvature representing the gravitational fields, as we will explain in what follows. Strictly speaking, it is actually four-dimensional spacetime that is curved, not three-dimensional space. And instead of assuming that gravitational forces make objects follow curved trajectories, Einstein assumed that they follow lines that are "as straight as possible" in these curved spacetimes – so-called "geodesics". As a result, the trajectories of the marshmallow and bowling ball do not depend on their masses at all. There is no need to make assumptions about the equivalence between gravitational and inertial mass; instead, the fact that they fall at the same rate is an immediate consequence of the properties of gravity. In the following subsections we will develop these concepts in some more detail, retracing our development of Newton's gravity in Section 1.1.

1.2.1 The Metric

Our first question should be: how can we measure the curvature of a space or even a spacetime? In a nutshell, we can measure curvature by measuring lengths. For example, we could measure the length C of a line of constant latitude on the Earth, and then measure the distance S from the pole to that latitude, as illustrated in Fig. 1.2. If the Earth were flat, the two numbers would be related by $C = 2\pi S$, but we would find $C < 2\pi S$ – a tell-tale sign that the surface of the Earth is curved.[5]

---

[4] Assuming that we can neglect friction, of course.
[5] An alternative would be to measure the sum of the interior angles in a triangle – in flat space, they will add to $180°$ but in general they will not. The interior angles in a triangle drawn on a sphere, for example, will add to more than $180°$. In the 1820s, long before the development of

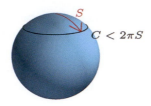

Figure 1.2 Illustration of one approach to measure curvature. The circumference of a circle of constant latitude on a sphere, C, is less than $2\pi S$, where S is the distance to the sphere's pole.

Now we need a tool for measuring lengths. Imagine that our space, or spacetime, is covered by coordinates, so that every point in space, or every event in the spacetime, is labeled by certain values of these coordinates. *A priori*, these coordinates need not have any physical meaning. Street addresses, for example, are very useful coordinates, and yet the difference in street number between your home and your neighbor's has nothing to do with the physical distance to your neighbor. To translate a difference in coordinate value into a physical distance we need a *metric*. In a one-dimensional space, the metric would simply be a factor that multiplies the "coordinate distance" so that it becomes a physical distance. In higher dimensions we need to employ something similar to a Pythagorean theorem. In general, we write the *line element* between two points whose coordinate labels differ by dx^a as

$$ds^2 = g_{ab}dx^a dx^b, \tag{1.18}$$

where g_{ab} is the *metric tensor*. Here the indices a and b run from 0 to 3, with 0 denoting time, and again we are using the Einstein summation convention by which we sum over repeated indices. The *invariance* of the line element under coordinate transformations, i.e. the fact that it takes the same value independently of the coordinate system in which it is evaluated, plays a key role in both special and general relativity.

We will consider the metric to be the fundamental quantity of general relativity; its role in Einstein's gravity is directly related to the role of the gravitational potential Φ in Newton's gravity. Note that the metric is a rank-2 tensor – i.e. it is a tensor of a rank that is two higher than its Newtonian counterpart. It has two indices, and we could display

general relativity, Carl Friedrich Gauss carefully measured the interior angles in a triangle defined by three mountain tops around Göttingen, Germany, in an attempt to ascertain the geometry of space. To within the accuracy of his measurements, he could not detect a deviation from $180°$.

the metric as a matrix (see below). We can already anticipate, then, that the relativistic cousins of all other objects that we encountered in Newton's gravity will likewise carry two more indices than their Newtonian counterparts. In particular, the relativistic generalization of the Newtonian field equation (1.15) will be a rank-2 tensor equation, as we will see in Section 1.2.4 below.

Readers with a knowledge of special relativity will already be familiar with one particular metric, namely the *Minkowski metric*. The Minkowski metric describes a *flat spacetime*; we often denote it by η_{ab}, and in Cartesian coordinates it takes the form[6]

$$g_{ab} = \eta_{ab} = \begin{pmatrix} -1 & 0 & 0 & 0 \\ 0 & 1 & 0 & 0 \\ 0 & 0 & 1 & 0 \\ 0 & 0 & 0 & 1 \end{pmatrix}. \tag{1.19}$$

As we noted above, the indices a and b now run over four coordinates, e.g. t, x, y, and z. We also note that we have assumed units here in which the speed of light is unity, $c = 1$. This may appear confusing, but, in fact, we informally do something very similar all the time. When asked, for example, how far Portland, Maine, is from Bowdoin College, the answer will often be "about half an hour", even though "hour" is not a unit of distance. The answer nevertheless makes sense, because there is a typical speed at which one travels between Bowdoin and Portland, namely the speed limit of 60 miles per hour or thereabouts. Similarly, we often refer to astronomical distances in units of "light-years"; a light-year is not a time but the distance that light travels in a year. We can therefore express distances in units of time. Here we will do the reverse and express time in units of distance. We again have a typical speed, namely the speed of light, and we will express time and space in units of length such that $c = 1$. In fact, henceforth we will express mass in the same units of length also, thereby setting $G = 1$. We refer to this arrangement as *geometrized units*. In these geometrized units, length, time, and mass all have the same unit, and combinations such as M/R are dimensionless. We can always recover expressions in physical units by reinserting appropriate powers of the constants G and c; we will practice that in the discussion around equation (1.70) and Exercise 1.12 below.

---

[6] Different authors adopt different sign conventions. We use the convention usually adopted by relativists in which space components are positive and time components are negative, but among particle physicists, for example, the opposite convention is also common.

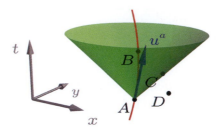

Figure 1.3 A spacetime diagram: time points "up" and the three-dimensional space is represented by the xy plane (where the z-dimension is suppressed). Light rays emitted from the event A form the light-cone represented by the green conical surface, and intervals between A and events on the light-cone, e.g. C, are light-like. Intervals inside the light-cone, e.g. between events A and B, are time-like, while those outside the light-cone, e.g. between A and D, are space-like. The red line represents a hypothetical worldline of a time-like observer; events on such a worldline must always be separated by time-like intervals. The observer's four-velocity u^a, which we introduce in equation (1.26) below, is tangent to the worldline.

Inserting the metric (1.19) together with the Cartesian displacement vector $dx^a = (dt, dx, dy, dz)$ into the line element (1.18) we obtain

$$ds^2 = -dt^2 + dx^2 + dy^2 + dz^2. \qquad (1.20)$$

As promised, equation (1.20) is the generalization of the Pythagorean theorem for flat spacetimes, where ds is the invariant interval (i.e. the distance in spacetime) between neighboring points (t, x, y, z) and $(t + dt, x + dx, y + dy, z + dz)$.

Since the "space" and "time" components of the spacetime metric enter with different signs, we can distinguish three different types of intervals between two nearby points in spacetime, i.e. two nearby *events*. *Space-like* intervals are those with $ds^2 > 0$, for which the separation in space dominates; for such intervals we define the infinitesimal proper distance dl as $dl = (ds^2)^{1/2}$. In the *spacetime diagram* of Fig. 1.3, for example, events A and D are separated by a space-like interval.

Time-like intervals are those with $ds^2 < 0$, for which the separation in time dominates; events A and B in Fig. 1.3 are separated by such a time-like interval. Space-like intervals can be measured by rods, time-like intervals by clocks. Since objects with non-zero mass always travel at speeds less than the speed of light, two events on such an object's trajectory must always be separated by a time-like interval. For such intervals we then compute the object's advance of, or infinitesimal change in, *proper time*, i.e. time as measured by a clock moving with the object, $d\tau = (-ds^2)^{1/2}$. Events separated by time-like intervals are *causally connected* because information can be sent from one to the other; events separated by space-like intervals are *causally disconnected*.

Finally, *light-like* or *null* intervals are those for which $ds^2 = 0$, e.g. the interval between events A and C in Fig. 1.3. In such a spacetime diagram, light-like intervals appear at an angle of 45°. Light, and potentially other particles that travel at the speed of light, travel along trajectories whose events are separated by light-like intervals. All such trajectories emitted from one point form the *light cone* of that point. Shown in Fig. 1.3 is the *future light cone* for light emanating from the event A and propagating to times $t > 0$. We have omitted the *past light cone*, an "upside down" cone with its apex at A, showing light rays emitted from $t < 0$ that arrive at A.

We can also transform the Cartesian form (1.19) of the Minkowski metric into other coordinate systems. In spherical polar coordinates, with coordinates t, R, θ, and φ, it takes the form

$$\eta_{ab} = \begin{pmatrix} -1 & 0 & 0 & 0 \\ 0 & 1 & 0 & 0 \\ 0 & 0 & R^2 & 0 \\ 0 & 0 & 0 & R^2 \sin^2 \theta \end{pmatrix}, \tag{1.21}$$

and the line element (1.18) becomes

$$\begin{aligned} ds^2 &= -dt^2 + dR^2 + R^2 d\theta^2 + R^2 \sin^2 \theta \, d\varphi^2 \\ &= -dt^2 + dR^2 + R^2 d\Omega^2, \end{aligned} \tag{1.22}$$

where we have introduced the *solid angle* $d\Omega^2 \equiv d\theta^2 + \sin^2 \theta d\varphi^2$.

As an example, we can measure the proper distance between the origin of the Minkowski spacetime to a point with radial coordinate R. We will choose a path at constant time t and angles θ and φ, so that we have $dt = d\theta = d\varphi = 0$. From (1.22) we then have $dl = (ds^2)^{1/2} = dR$, and evidently

$$l = \int_0^R dl = \int_0^R dR' = R, \tag{1.23}$$

which is as expected.

Exercise 1.2 Consider a circle in the equatorial plane (i.e. $\theta = \pi/2$) of the Minkowski spacetime, at constant time t and constant coordinates R and θ. Confirm that this circle has proper length $C = 2\pi R$. We hence conclude that the Minkowski spacetime is flat.

It is time to place the notion of a "vector" into the more general arena of "tensors". We summarize the basic properties of tensors in Appendix A and invite the reader unfamiliar with them to read Sections A.1 and A.2 before proceeding further. There we make clear the difference between tensor components with indices "upstairs" and with indices "downstairs", e.g. A^a versus A_a. Technically, we refer to upstairs components as *contravariant* components, and downstairs components as *covariant*.[7] More precisely, the contravariant components appear in the expansion of a tensor in terms of basis *vectors* ("arrows"), while the covariant components appear in the expansion in terms of basis *one-forms* ("surfaces"). Vectors and one-forms are both tensors of rank 1 (their components have one index). In general, rank-n tensors have components with n indices. As we discuss in more detail in Appendix A, basis vectors and basis one-forms, and hence the two types of components, differ in the way in which they transform in coordinate transformations. The generic example of a vector is the displacement vector with components dx^a, which we encountered above, and that of a one-form is the gradient, with components $\partial_a f$ where f is a scalar function. Vectors and one-forms transform in "opposite" ways (see equations A.28 and A.29 below). In a dot product, e.g. $A_a B^a$, we sum over one contravariant and one covariant component, so that the two different transformations cancel each other out and the result remains invariant – as it should for a scalar.

We can always relate a contravariant component to a covariant component by "lowering" the index with the help of the metric, e.g. $A_a = g_{ab} A^b$. In particular, this allows us to write the dot product between two rank-1 tensors as $\mathbf{A} \cdot \mathbf{B} = A_a B^a = g_{ab} A^b B^a$. When all the indices appear in pairs, one upstairs and one downstairs, we know we have formed an invariant. We can compute the magnitude of a vector from the dot product of the vector with itself; for the displacement vector dx^a, for example, we obtain the line element (1.20). Similarly to our classification of spacetime intervals above we can now distinguish space-like, time-like, and light-like (or null) vectors.

---

[7] Unfortunately the word "covariant" has two different meanings: sometimes we use it to describe objects that properly transform as tensors between different coordinate systems, other times to refer to downstairs components. While this is potentially confusing, the meaning should be clear from the context.

Exercise 1.3 (a) Show that the two vectors $A^a = (-1,2,0,0)$ and $B^a = (-2,1,0,0)$ are orthogonal in a Minkowski spacetime, i.e. $A_a B^a = 0$.

(b) Draw the two vectors in a spacetime diagram, and note that they do not make a "right angle" as they do in Euclidean spaces.

We can also "raise" an index with the inverse metric, e.g. $A^a = g^{ab} A_b$. The inverse metric is defined by the requirement that $g^{ab} g_{bc} = \delta^a{}_c$, where $\delta^a{}_c$ is the Kronecker delta symbol, which we previously encountered in Exercise 1.1: it equals one when the index a is the same as c and is zero otherwise (see equation A.7). In practice, we compute the components of the inverse metric exactly as we compute the inverse of a matrix in linear algebra.

Finally we return to a subtlety in notation that we quietly introduced in Section 1.1. Consider equation (1.4), where we used boldface notation to denote the vectors on the left but used their components on the right (e.g. $\mathbf{a}$ versus a_i). Now, the value of a component of a tensor applies only in a *specific* coordinate system. For example, equation (1.19) provides the components of the metric for a flat spacetime in Cartesian coordinates. Clearly this relation holds only in that particular coordinate system (in spherical polar coordinates, for instance, the metric takes the form 1.21). However, when we write down equations relating tensor components on both sides of the equation ("tensor equations") they hold in *any* coordinate system. An example might be $A^a = B^a$; i.e. the components A^a of vector $\mathbf{A}$ are equal to the components B^a of vector $\mathbf{B}$ in any coordinate system. Given that the two sets of components are always identical, this equation represents a relation between the tensors themselves. This suggests an *abstract tensor notation*, by which A^a, for example, represents the vector itself rather than its individual components. Likewise, $A^a = B^a$ is no longer a relation between tensor components but is instead a coordinate-independent tensor equation. We will use this notation frequently in this book. In many cases this notation is much clearer than boldface, since it allows for an unambiguous specification of the rank and type of the tensor, by means of sub- or superscripts.

Returning to our development of general relativity, recall that we started this section by asking how we can measure the curvature of space or spacetime. As a first tool we introduced the metric – but now we observe that the metrics (1.19) and (1.21) look different even though they describe exactly the same (flat) spacetime. How can we then measure curvature in a coordinate-independent way? Doing that requires the tools of differential geometry. We will sketch how these tools can be developed, beginning with some of the most important concepts in the following sections. A partial preview and summary of the next section is contained in Appendix A.3.

1.2.2 The Geodesic Equation and the Covariant Derivative

As we discussed above, Einstein proposed that freely falling particles, both with and without mass, follow the shortest possible path through (potentially) curved spacetimes. By "freely falling" particles we mean particles on which no forces act, but we note that this does not include gravity – gravity is accounted for by the curvature of spacetime, not by a force. For light we already know this to be true, since it is in accordance with Fermat's principle.[8] Essentially, Einstein extended Fermat's principle to objects with mass. Such objects travel at speeds less than the speed of light, so that we can measure their proper time. For these objects, the trajectory of shortest possible path *maximizes* the advance of the object's proper time. This sounds confusing, but readers familiar with special relativity will recognize this fact from the so-called twin paradox: one twin stays on the Earth, while the other twin travels at high speed to a galaxy far, far away and then returns. Upon return the twin who stayed on the Earth has aged more than the traveling twin. The asymmetry between the aging appears paradoxical until one recognizes that, in order to return to Earth, the traveling twin had to turn around. This requires acceleration, meaning that that twin was not freely falling at all times, while the homebound one was (at least approximately). The twin on Earth ages more than any traveling twin, in accordance with the proposition that a freely falling trajectory will maximize the advance of proper time.

When we draw the trajectory of an object through three-dimensional space, we select different points, each of which represents the location of the object at one instant of time. We then connect these points to form a line. At each point we can also draw a tangent to the line; this tangent represents the object's velocity at the corresponding instant of time:

$$v^i = \frac{dx^i}{dt}. \tag{1.24}$$

---

[8] Fermat's principle states that the path taken by a ray of light between two given points is the path that can be traversed in the least time.

In the absence of any forces an object travels in a straight line and does not accelerate,

$$a^i = \frac{dv^i}{dt} = 0;$$ (1.25)

it therefore has a constant velocity vector that remains the same all along the object's trajectory.

In a four-dimensional spacetime diagram we include time as an additional axis. A point in a spacetime diagram represents an *event*; its coordinates give us its location both in space and in time. Connecting the different events of an object's trajectory yields the object's *worldline*. The red line in Fig. 1.3, for example, represents such a worldline. Tangents to this worldline now represent the *four-velocity*, defined as

$$u^a = \frac{dx^a}{d\tau},$$ (1.26)

where τ is the object's proper time.

Exercise 1.4 Show that the four-velocity is *normalized*,

$$u_a u^a = -1.$$ (1.27)

Hint: Recall how to compute the advance of proper time $d\tau$ along time-like intervals.

We can now generalize the notion of a "straight line" to curved spacetimes by constructing curves for which the derivative of the four-velocity u^a, taken in the direction of the four-velocity, vanishes,

$$a^a = \frac{Du^a}{D\tau} = u^b \nabla_b u^a = 0.$$ (1.28)

Here a^a is the four-acceleration and (1.28) is the *geodesic equation*. We again employ the Einstein summation convention in the third expression, and we also encounter a new type of derivative, namely the *covariant derivative*. In four dimensions we will denote the covariant derivative by the nabla symbol ∇. The symbol D in the second expression above reminds us that the time derivative is covariant. The notion of the covariant derivative is needed for the following reason.

For a scalar function we can unambiguously compare values of the function at two different points and from those values compute the derivative, representing the change in the function divided by the separation of points in the limit where the separation goes to zero.

That operation is what we do when we compute a partial derivative of a scalar function. For a vector or tensor, however, it is not good enough to compute the partial derivatives of the components when comparing a quantity at nearby points in curved spacetimes. Remember that a vector represents a linear combination of *basis vectors*, the coefficients of which are the *components* of the vector. When a vector has an r-component A^r, say, we can construct the vector **A** from the product $A^r \mathbf{e}_r$, plus similar terms for the other components (see equation A.1 in Appendix A). Therefore, when we compute the derivative of a vector **A**, we have to take into account not only changes in the components from one point to another but also changes in the basis vectors $\mathbf{e}_a$. Even in flat spacetime these basis vectors may change, unless one is employing Cartesian (rectangular) coordinates. These changes in basis vectors are encoded in the so-called *Christoffel symbols* $^{(4)}\Gamma^a_{bc}$; see equation (A.34). We decorate the Γs with a superscript $^{(4)}$ as a reminder that here we are referring to four-dimensional spacetime. We will later encounter their three-dimensional spatial counterparts, which are associated with the covariant derivative in three dimensions.

For a scalar quantity, the covariant derivative is the same as a partial derivative – because there are no basis vectors to consider. For a vector **V** we can express the covariant derivative as

$$\nabla_a V^b = \partial_a V^b + {}^{(4)}\Gamma^b_{ca} V^c. \tag{1.29}$$

Here the first term is responsible for the changes in the vector's components, and the second for the changes in the basis vectors. For higher-rank tensors there will be one additional Christoffel term for each additional index attached to the tensor. We usually learn to work with covariant derivatives and Christoffel symbols in the context of curved spaces in differential geometry or general relativity, but they are important even in flat spaces whenever curvilinear coordinates are used – this is why the divergence, curl, or Laplace operator are more complicated in spherical or cylindrical coordinates than in Cartesian coordinates; see exercise A.9 for an example.

We construct the Christoffel symbols in such a way that the covariant derivative measures the deviation from *parallel transport*. In curved spaces it is impossible to decide whether two vectors, residing at two different points, are parallel. However, we can move a vector along some path from one point to another, keeping it "as parallel as possible" locally. Consider, for example, a vector field V^a that is tangent to

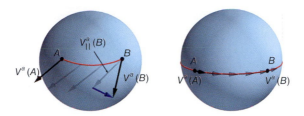

Figure 1.4 Left-hand panel: Parallel transport and covariant derivative. Consider a vector field $V^a(x^b)$, tangent to the sphere, that varies from point to point. The solid black arrows show two such vectors, $V^a(A)$ and $V^a(B)$, tangent at points A and B respectively. Pick up the tangent vector $V^a(A)$ and move it to B along the red curve, keeping it as parallel to itself as possible, while also keeping it tangent to the sphere. This defines parallel-transported vectors $V^a_{||}$ (light gray). The covariant derivative of V^a along the red curve measures the difference (shown as the blue vector) between $V^a(B)$ and the parallel-transported vector $V^a_{||}(B)$ in the limit that the displacement from A to B goes to zero. Right-hand panel: A geodesic. Pick up the vector $V^a(A)$ at A, parallel-transport it along itself, as shown. The curve generated by this process is a geodesic curve and the covariant derivative of V^a along it is zero, by construction. On a sphere this generates a great circle.

a sphere, as illustrated in the left-hand panel of Fig. 1.4. Parallel-transporting the vector $V^a(A)$ from the point A to the point B along the red curve yields the light gray vectors $V^a_{||}$. If a second vector field, T^a, is tangent to the red curve, we say "$V^a_{||}$ is parallel-transported along T^a". The covariant derivative then measures the difference between the vector $V^a(B)$ and the parallel-transported vector $V^a_{||}(B)$, shown as the blue vector, in the limit that the displacement from A to B goes to zero. Evidently, the covariant derivative of the parallel-transported field $V^a_{||}$ along the tangent vector T^a vanishes by construction,

$$T^a \nabla_a V^b_{||} = 0. \tag{1.30}$$

For a geodesic, the tangent vector to a curve is parallel-transported along itself – i.e. it remains as parallel as possible, as illustrated in the right-hand panel of Fig. 1.4. For an object with non-zero rest mass traveling through spacetime along its worldline, the tangent to the worldline is the four-velocity u^a, and the geodesic equation takes the form (1.28), which states that the four-acceleration of such an object is zero.

For now it is sufficient to recognize that the Christoffel symbols constructed in this way contain first derivatives of the metric, but for completeness we will provide a mathematical expression appropriate for any coordinate basis:

$$^{(4)}\Gamma^a_{bc} = \frac{1}{2}g^{ad}(\partial_c g_{db} + \partial_b g_{dc} - \partial_d g_{bc}) \qquad (1.31)$$

(recall the summation over repeated indices; see Section A.3 in Appendix A for more details). Since the Christoffel symbols are computed from first derivatives of the metric, it is not surprising that they have three indices – two for the metric, and one for the first derivative. Extending the analogy between the metric g_{ab} and the Newtonian potential Φ as the fundamental quantities of Einstein's and Newton's theories, respectively, we anticipate that a combination of Christoffel symbols plays the same role in the former theory as the gravitational fields $-D_i\Phi$ play in the latter theory. As we expected, they have two more indices than their Newtonian counterparts.

Expression (1.29) holds for the contravariant component of the covariant derivative; the covariant component of a covariant derivative is given by (A.36). Exercise A.6 shows that the covariant derivative of a scalar product $A_a B^a$ reduces to an ordinary partial derivative, as one would expect. For a rank-n tensor, the covariant derivative has n Christoffel-symbol terms: one with a plus sign as in (1.29) for each contravariant component, and one with a negative sign as in (D.5) in Appendix D for each covariant component.

Returning to our narrative, we insert (1.29) into (1.28) and notice the similarity of the relativistic equations of motion, the geodesics (1.28), to their Newtonian counterparts (1.6): the derivative of the four-velocity u^a is an acceleration, and expanding the covariant derivative introduces the Christoffel symbols, which, in turn, are related to the Newtonian gravitational fields embedded in $-D_i\Phi$. In Section 1.1 we observed that the first derivatives $D_i\Phi$ do not provide a frame-independent measure of the gravitational fields, and the reader will not be shocked to hear that neither do the Christoffel symbols $^{(4)}\Gamma^a_{bc}$.

1.2.3 Geodesic Deviation and the Riemann Tensor

Following our Newtonian roadmap we now reach item 3 in the list at the end of Section 1.1, suggesting that we consider the deviation between two nearby freely falling objects in order to obtain a frame-independent measure of the gravitational fields, i.e. the curvature of spacetime. Conceptually we will follow the same steps as those we used in our Newtonian development: we evaluate the equations of motion at two nearby points, use a Taylor expansion, subtract the two results, and obtain an equation for the deviations Δx_i. The calculation is more complicated in this case, however, both because the relativistic equations of motion (1.28) are more complicated than their Newtonian counterpart (1.6) and also because we no longer want to restrict our analysis to Cartesian coordinates.

Instead of getting caught up in those details, we will observe the following. The geodesic equation (1.28) contains up to first derivatives of our fundamental quantity, the metric g_{ab}, well hidden in the Christoffel symbols $^{(4)}\Gamma^a_{bc}$. Using a Taylor expansion to derive the deviation between two geodesics will introduce second derivatives of g_{ab}, in exactly the same way as the Taylor expansion (1.9) introduced second derivatives of Φ in (1.12). Furthermore, just as in the Newtonian case, where we defined the spatial "tidal tensor" $\mathcal{R}_{ij}$ in (1.11) to absorb the second derivatives of Φ, we can introduce a tensor that absorbs the second derivatives of g_{ab} together with a few nonlinear terms. This tensor is the famous *Riemann tensor* $^{(4)}R^a_{bcd}$. As in the case of the Christoffel symbols, we will decorate the spacetime Riemann tensor with a superscript $^{(4)}$ in order to distinguish it from its three-dimensional, spatial counterpart R^i_{jkl}, which we will meet in Chapter 2. With the help of this Riemann tensor the relativistic equation of the geodesic deviation becomes

$$\frac{d^2 \Delta x^a}{d\tau^2} = -^{(4)}R^a_{bcd} u^b u^d \Delta x^c, \tag{1.32}$$

and we notice the similarity to its Newtonian analogue (1.12).

We refer to Appendix D.2 for the mathematical expressions for and properties of the Riemann tensor. In particular, the Riemann tensor satisfies the symmetries (D.12), and its derivatives are related by the *Bianchi identity* (D.13), which will play an important role in Chapter 2. For our purposes here, however, the Riemann tensor's most important characteristic is that it contains second derivatives of the metric g_{ab}. Given that the latter, as a rank-2 tensor, already has two indices, it is not surprising that the Riemann tensor has four indices, making it a rank-4 tensor. As we anticipated, it has two more indices than its Newtonian counterpart, the tidal tensor $\mathcal{R}_{ij}$, but plays the same role as the Newtonian tidal tensor: it is responsible for tides and provides a frame-independent measure of the gravitational field, which we now describe as the *spacetime curvature*.

Recall that, in the context of Newtonian gravity, the gravitational forces themselves are not frame-independent and disappear in a freely falling frame of reference. The tidal tensor, however, provides a frame-independent measure of the gravitational fields. The same is true in general relativity. It can be shown that we can always choose coordinates in which the first derivatives of the metric, and hence the Christoffel symbols (1.31), vanish, and *locally* the metric itself becomes the Minkowski metric (1.19). We refer to such a "freely falling" coordinate choice as a *local Lorentz frame*. Even in such a local Lorentz frame, the Riemann tensor, like the Newtonian tidal tensor, will not in general be zero. For a flat spacetime, however, when the metric can be expressed by the Minkowski metric *globally*, the Riemann tensor vanishes identically, independently of the coordinates in which we evaluate it.

We have introduced the Riemann tensor using the notion of geodesic deviation, but an alternative motivation is based on the concept of parallel transport. Consider parallel-transporting a vector V^a around a closed loop. It is intuitive that, in flat spaces, we always end up with the same vector that we started with. This is not true in curved spaces! As a concrete example, start with a vector V^a pointing north at some point on the equator of the Earth, as shown in Fig. 1.5. After parallel-transporting V^a along the equator halfway around the Earth it will still point

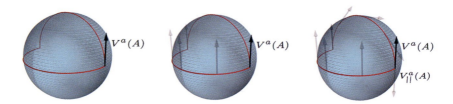

Figure 1.5 Illustration of curvature via parallel transport around a closed loop. The vector V^a is parallel-transported around the loop marked by the red curve. Left-hand panel: We start with a vector V^a pointing north at some location A on the equator. Middle panel: We parallel-transport the vector halfway around the Earth along the equator. Right-hand panel: Now we parallel-transport V^a back to the original location, but this time via the north pole. Arriving at the original location, the parallel-transported vector $V^a_{\parallel}$ now points south, not north, i.e. its direction has rotated by 180°. In order to verify that the amount by which the vector rotates is related to the curvature inside the loop, convince yourself that the vector will turn by only 90° if you make the loop half as large, for example by moving the vector only a quarter of the way around the Earth along the equator, then to the north pole, then back to the original location.

north. Now we parallel-transport V^a to the north pole, and continue to our original starting point – where it will arrive pointing south! The difference between a vector V^a and its copy after parallel transport along a closed loop therefore provides a measure of the curvature inside the loop – similar to the ratio between radius and circumference that we have invoked before. Computing this difference for an infinitesimal loop leads to the definition (D.10) of the Riemann tensor.

1.2.4 Einstein's Field Equations

Our Newtonian roadmap provides guidance for the next step as well, namely, how to find the field equations that determine the metric g_{ab}. As we saw in Section 1.1, the Newtonian field equation (1.15), also known as the Poisson equation, relates the trace of the tidal tensor to the rest-mass density; by analogy the relativistic field equations, Einstein's equations, should relate the trace of the Riemann tensor to some measure of the total mass–energy density.

Forming the trace of a tensor means summing over a pair of its indices. The Riemann tensor has four indices, so it appears as if there could be multiple choices for computing this trace. However, because of the symmetries of the Riemann tensor (see Appendix D.2), there is only one meaningful way of computing its trace to get a rank-2 tensor, so that, up to a sign, there is no ambiguity. We call this trace the *Ricci tensor*,

$$^{(4)}R_{ab} \equiv {}^{(4)}R^c{}_{acb}, \tag{1.33}$$

where the index c is summed over. The result is a rank-2 tensor, as we expect from our comparisons with Newton's theory. We can express the Ricci tensor in a number of different ways, either involving the first derivatives of the Christoffel symbols or the second derivatives of the metric, as shown in equations (D.16)–(D.18) of Appendix D. Since the Ricci tensor is so important for our discussion in the following we list one particularly useful expression here as well,

$$^{(4)}R_{ab} = \frac{1}{2}g^{cd}\left(\partial_a\partial_d g_{cb} + \partial_c\partial_b g_{ad} - \partial_a\partial_b g_{cd} - \partial_c\partial_d g_{ab}\right)$$
$$+ g^{cd}\left({}^{(4)}\Gamma^e_{ad}\,{}^{(4)}\Gamma_{ecb} - {}^{(4)}\Gamma^e_{ab}\,{}^{(4)}\Gamma_{ecd}\right), \tag{1.34}$$

where $^{(4)}\Gamma_{abc} = g_{ad}{}^{(4)}\Gamma^d_{bc}$. At the same time we also define the *Ricci scalar* as the trace of the Ricci tensor,

$$^{(4)}R \equiv {}^{(4)}R^a{}_a. \tag{1.35}$$

Here we have raised the first index of $^{(4)}R_{ab}$ by contracting with the inverse metric, $^{(4)}R^a{}_c = g^{ab\,(4)}R_{bc}$; see Appendix A.

An expression involving $^{(4)}R_{ab}$ and $^{(4)}R$ appears on the left-hand side of Einstein's field equations. Let us now discuss the right-hand side of the field equations. The Newtonian version, equation (1.15), features the rest-mass density ρ_0 on the right-hand side. The relativistic generalization is the rank-2 *stress–energy* tensor T^{ab}. For different types of mass–energy sources (baryons, fluids, electromagnetic fields, etc.) the stress–energy tensor depends on these sources in different functional forms, but we can always interpret the time–time component T^{tt} as the total mass–energy density (e.g. ρ for baryons), the time–space components T^{ti} as an energy flux or momentum density, and the purely spatial components T^{ij} as a momentum flux or stress tensor (e.g. $T_{ij} = P\delta_{ij}$ in isotropic fluids, where P is the pressure). In vacuum we have, not surprisingly, $T^{ab} = 0$.[9]

We have already suspected that the Ricci tensor $^{(4)}R_{ab}$ should make an appearance on the left-hand side of Einstein's equations, but other candidates for curvature-related rank-2 tensors are $^{(4)}Rg_{ab}$ and the metric g_{ab} itself. It turns out that $^{(4)}R_{ab}$ and $^{(4)}Rg_{ab}$ can appear only in a certain combination, since otherwise energy and momentum would not be conserved.[10] We therefore absorb these two terms in the *Einstein tensor*,

$$G_{ab} \equiv {}^{(4)}R_{ab} - \frac{1}{2}{}^{(4)}R\,g_{ab}. \tag{1.36}$$

Further requiring that our field equations reduce to the Newtonian field equation (1.15) in the Newtonian limit determines the relation between the Einstein tensor G_{ab} and the stress–energy tensor T_{ab}, leaving us with

$$\boxed{G_{ab} + \Lambda g_{ab} = 8\pi T_{ab}.} \tag{1.37}$$

---

[9] Note, however, that the absense of matter does not mean that spacetime is flat: both black holes and gravitational waves are vacuum solutions; see Section 1.3.

[10] As we will discuss in greater detail in Section 2.3, the divergence of G_{ab} is identically zero for all spacetimes by construction, as is the case for g_{ab}. By virtue of Einstein's equations (1.37) below this implies that the divergence of the stress–energy tensor T_{ab} must also vanish. The latter gives rise to equations of motion for the matter sources, which in turn guarantee conservation of their energy and momentum.

This is our jewel, truly deserving a box – *Einstein's field equations of general relativity.*[11]

The term Λ in Einstein's equations (1.37) is the famous *cosmological constant*. Einstein introduced this term in order to reproduce a static, time-independent, universe but quickly abandoned it after Edwin Hubble discovered the expansion of the universe.[12] Ironically, the recently discovered acceleration of the universe's expansion can be described with the help of such a cosmological constant, where it is often interpreted as a "dark energy" component of the universe whose stress–energy tensor mimics this cosmological constant term. From observations, the value of Λ is sufficiently small that it has significant effects only on large cosmological scales. Therefore, Λ is irrelevant for most non-cosmological applications in numerical relativity; accordingly we will disregard this term in what follows.

Note an aesthetic feature of Einstein's equations: the geometry is on the left-hand side, and mass–energy on the right-hand side. This leads to the slogan, coined by John Wheeler: "Spacetime tells matter how to move; matter tells spacetime how to curve".

Einstein's equations (1.37) form partial differential equations for the metric g_{ab}, just like its Newtonian counterpart, the Poisson equation (1.15), is a partial differential equation for the Newtonian potential Φ. However, Einstein's equations are significantly more complicated than the Poisson equation: in four-dimensional spacetime there are ten equations (the tensors in 1.37 are symmetric in their indices a and b), and they are nonlinear in the gravitational field variables. As a consequence, analytical solutions to Einstein's equations have been found typically by imposing a number of simplifying assumptions, e.g. symmetries. One example is the general spacetime solution for a stationary black hole, which we will discuss in Section 1.3.1 below. Finding general solutions for most cases of physical interest – for example, calculating the inspiral and merger of two black holes, together with the emitted gravitational wave signal – requires solving the equations via alternative methods. In this volume we will discuss numerical relativity as one such

---

[11] See Einstein (1915), which did not yet include the term Λ. Einstein's equations can also be derived from a variational principle using the geometric Lagrangian density $\mathcal{L}_{geom} = R/(16\pi)$ and a matter Lagrangian density $\mathcal{L}_{mat}$. Varying the Hilbert action $S = \int (\mathcal{L}_{geom} + \mathcal{L}_{mat})\sqrt{-g}\,d^4x$ yields Einstein's equations $G_{ab} = 8\pi T_{ab}$, where $T_{ab} = -2\delta\mathcal{L}_{mat}/\delta g^{ab} + g_{ab}\mathcal{L}_{mat}$ (see Hilbert (1915), also Misner et al. (2017); Poisson (2004)).

[12] It is often reported that Einstein considered the introduction of the cosmological constant his "biggest blunder".

method – that will be the focus of the following chapters. Before we do that, though, we should review at least two important analytical results.

1.3 Two Important Analytical Solutions

1.3.1 Schwarzschild Black Holes

Shortly after Einstein put his final touches to his theory of general relativity, Karl Schwarzschild[13] famously derived an exact, static, spherically symmetric vacuum solution for the metric,

$$ds^2 = -\left(1 - \frac{2M}{R}\right) dt^2 + \left(1 - \frac{2M}{R}\right)^{-1} dR^2 + R^2 d\Omega^2. \quad (1.38)$$

Tragically, Schwarzschild died only months later, while serving in World War I, decades before this solution was recognized as describing non-rotating black holes and fully appreciated for its monumental astrophysical importance.

The constant M in (1.38) turns out to be the black hole's mass. In particular we see that, for $M = 0$, we recover the flat Minkowski metric (1.21), which is reassuring. *A priori* the choice of a radial coordinate to describe the spacetime need not have any particular physical meaning (think of street addresses), but it turns out that in this case R does. We could repeat exercise 1.2 for the line element (1.38) to find that the proper circumference C of a circle at radius R is $2\pi R$, as we might expect from our flat-space intuition. Similarly, a sphere of radius R has a proper area $A = 4\pi R^2$, again in accordance with our intuition. Both the proper circumference and proper area are scalar invariants – all coordinate choices would yield the same values. We therefore refer to the radius R that appears in the Schwarzschild line element (1.38) as the *circumferential* or *areal* radius and note that it can be determined from measurable scalar invariants: $R = C/(2\pi) = (A/(4\pi))^{1/2}$.

The Schwarzschild spacetime is not flat. In fact, computing the Riemann tensor $^{(4)}R^a_{bcd}$ would show that the curvature of the spacetime diverges as we approach $R \to 0$, meaning that the center

---

[13] See Schwarzschild (1916); see also Droste (1917) for another early and independent derivation.

of the Schwarzschild spacetime features a *curvature* or *spacetime singularity*.[14] For large separations, $R \gg M$, however, we recover the flat Minkowski metric (1.21). We refer to spacetimes with this property for large R as *asymptotically flat*.

It would be wonderful if we had a way to visualize the curvature of the Schwarzschild spacetime. Generally, when we visualize curvature, we imagine the shape that an object would take in a higher-dimensional space. Sometimes we refer to this as *embedding* the object in a higher-dimensional space. For example, the curvature of a two-dimensional sphere is quite intuitive to us, because we are familiar with the shape that such a sphere takes in three dimensions. Visualizing the curvature of the four-dimensional Schwarzschild spacetime, on the other hand, would require even higher dimensions – that is not really an option. The best we can do, therefore, is to visualize a subspace of the Schwarzschild spacetime.

We start by focusing on one instant of time, $t = const$. In Chapter 2 we will refer to the remaining subspace as a "spatial slice" of our spacetime. We then restrict attention to the equatorial plane, i.e. we choose $\theta = \pi/2$, which leaves us with the two-dimensional line element

$$^{(2)}ds^2 = \left(1 - \frac{2M}{R}\right)^{-1} dR^2 + R^2 d\varphi^2. \tag{1.39}$$

We can now visualize the curvature of this space with the help of a curved surface in a flat three-dimensional space as follows. We start by writing the line element of the flat three-dimensional space in cylindrical coordinates,

$$^{(3)}ds^2 = dR^2 + R^2 d\varphi^2 + dz^2. \tag{1.40}$$

Next we construct a two-dimensional surface whose "height" $z = z(R)$ depends on R only. Defining $z' = dz/dR$, we can find distances along this surface from

$$^{(2)}ds^2 = \left(1 + (z')^2\right) dR^2 + R^2 d\varphi^2. \tag{1.41}$$

We now choose the height $z(R)$ of the surface in such a way that the distance between points at different values of R and φ measured along the surface is the same as the proper distance in the space (1.39).

---

[14] Unlike the coordinate singularity at $R = 2M$, this curvature singularity cannot be removed by a coordinate transformation; see the discussion below.

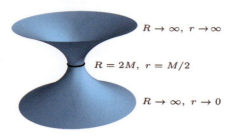

$R \to \infty, \; r \to \infty$

$R = 2M, \; r = M/2$

$R \to \infty, \; r \to 0$

Figure 1.6 Embedding diagram for a $t = const$ slice of the Schwarzschild solution (1.38). The black ring marks the event horizon at $R = 2M$. The labels r refer to the isotropic radius introduced in exercise 1.5. The diagram connects two asymptotically flat regions, one for $r = \infty$ and the other for $r = 0$, which we can think of as two separate universes. The two universes are connected by the "wormhole" shown in the figure.

For points on a surface of constant R, i.e. points found by varying the angle φ only, this is automatically the case – a consequence of R being the circumferential radius in (1.38). For points at different radii R, however, we have to equate the corresponding coefficients of dR in (1.39) and (1.41),

$$\left(1 - \frac{2M}{R}\right)^{-1} = 1 + (z')^2, \qquad (1.42)$$

and solve the resulting differential equation for z. Solving for z' and integrating we find

$$z = \pm(2M)^{1/2} \int \frac{dR}{(R - 2M)^{1/2}} = \pm(8M(R - 2M))^{1/2}, \qquad (1.43)$$

where we have omitted an irrelevant constant of integration. In an *embedding diagram* we graph z as a function of R and φ, as shown in Fig. 1.6. The curvature of this surface now encapsulates the curvature of the $t = const, \theta = \pi/2$ subspace of the Schwarzschild geometry (1.38). The key is that proper distances between points are measured by staying on the two-surface, whose curvature reliably reflects the Schwarzschild geometry.

Note that we have no solutions for $R < 2M$ but two different branches of solutions for $R > 2M$. The two branches connect at $R = 2M$ and both are asymptotically flat, with $z' \to 0$ as $R \to \infty$. This behavior is a consequence of the fact that slices with $t = const$ in the Schwarzschild spacetime connect two asymptotically flat ends of the spacetime, which we can think of as two separate universes. As suggested by the shape of the embedding diagram in Fig. 1.6, we refer to the geometry of these $t = const$ surfaces as a *wormhole* geometry, the wormhole being the region that connects the two asymptotically flat universes at the top and bottom of the figure.

It can be shown that the Schwarzschild geometry (1.38) is unique for spherically symmetric vacuum spacetimes; this result is referred to as *Birkhoff*'s *theorem*. While the geometry is unique, the coordinate system in which it can be described is not. The following exercises provide examples of other useful coordinate systems for the Schwarzschild spacetime.

Exercise 1.5 Consider a new "isotropic" radius r that satisfies

$$R = r \left(1 + \frac{M}{2r}\right)^2. \tag{1.44}$$

(a) Show that under this transformation the Schwarzschild metric (1.38) takes the form

$$ds^2 = -\left(\frac{1 - M/(2r)}{1 + M/(2r)}\right)^2 dt^2 + \psi^4(dr^2 + r^2 d\Omega^2) \tag{1.45}$$

and find the "conformal factor" ψ. Note that the spatial part of this metric takes the form $\psi^4 \eta_{ij}$, where η_{ij} is the flat metric in any coordinate system (here, in spherical polar coordinates; in Cartesian coordinates $\eta_{ij} = \delta_{ij}$). This form of the spacetime metric is called "isotropic" since its spatial part does not single out any direction.

(b) Find the inverse expression for r in terms of R. Show that real solutions for r exist only for $R \geq 2M$.

Exercise 1.6 (a) Show that under the coordinate transformation

$$dt = d\bar{t} - \frac{1}{f_0} \frac{M}{R - M} dR \tag{1.46}$$

the Schwarzschild metric (1.38) takes the new form

$$ds^2 = -f_0 d\bar{t}^2 + \frac{2M}{R - M} d\bar{t} dR + \left(\frac{R}{R - M}\right)^2 dR^2 + R^2 d\Omega^2, \tag{1.47}$$

where $f_0 = 1 - 2M/R$.

(b) Now transform to a new spatial coordinate $\bar{r} = R - M$ and show that the new metric takes the form[15]

$$ds^2 = -\frac{\bar{r} - M}{\bar{r} + M} d\bar{t}^2 + 2 \frac{M}{\bar{r}} d\bar{t} d\bar{r} + \psi^4(d\bar{r}^2 + \bar{r}^2 d\Omega^2) \tag{1.48}$$

---

[15] See Dennison and Baumgarte (2014); see also Dennison et al. (2014) for a generalization for rotating black holes.

with

$$\psi = \left(1 + \frac{M}{\bar{r}}\right)^{1/2}. \tag{1.49}$$

Note that the spatial part of the metric is again isotropic, so $\bar{r}$ is again an isotropic radius, as in exercise 1.5. Here, however, r is measured on a different $\bar{t} = const$ time slice; therefore it is different from the isotropic radius r in exercise 1.5.

The defining property of black holes is the existence of a *horizon*. Different notions of horizons make this statement precise in different ways. *Event horizons* separate those regions of a spacetime from which photons can escape to infinity from those regions from which they cannot escape. The event horizon is the "surface" of a black hole. For the purposes of numerical relativity, the notion of an *apparent horizon* is often of greater practical importance. An apparent horizon is the outermost *trapped surface*, where a trapped surface is defined as a closed surface on which the expansion (spread) of an outgoing bundle of light rays is zero or negative. On the apparent horizon itself, which encloses the region of all trapped surfaces, the expansion is zero. Unlike the event horizon, an apparent horizon, if it exists,[16] can be located knowing the metric at a single instant of time. For static (Schwarzschild) and stationary (Kerr) black holes, these different horizons coincide, and we will simply refer to them as "the horizon". For a Schwarzschild black hole the horizon is a spherical surface whose location can be identified by locating the radius at which an outgoing radial photon no longer propagates to larger areal radii.

To locate the horizon for the metric (1.48), we consider radial photons, i.e. photons along whose trajectory only dr and dt are non-zero. The trajectory must also be light-like, so that $ds^2 = 0$. From (1.48) we then have

$$\left(1 + \frac{M}{\bar{r}}\right)^2 d\bar{r}^2 + 2\frac{M}{\bar{r}}d\bar{t}\,d\bar{r} - \frac{\bar{r} - M}{\bar{r} + M}d\bar{t}^2 = 0 \tag{1.50}$$

or

$$\frac{d\bar{r}}{d\bar{t}} = \left(\pm 1 - \frac{M}{\bar{r}}\right)\frac{\bar{r}^2}{(\bar{r} + M)^2}. \tag{1.51}$$

---

[16] For some coordinate choices, an apparent horizon may not be present even if an event horizon, and hence a black hole, forms.

Here the "+" sign yields the coordinate speed of outgoing photons, while the "−" sign yields that of ingoing photons. We should also emphasize that any local observer would still measure photons propagating at the speed of light; the coordinate speed that we compute here has no immediate physical meaning. For $\bar{r} \gg M$, the coordinate speeds (1.51) reduce to ± 1, consistent with the observation that, like the Schwarzschild metric (1.38), the metric (1.48) approaches the Minkowski metric (1.21) in that limit (recall that we are using units in which $c = 1$). At $\bar{r} = M$, however, we see that even "outgoing" photons no longer propagate to larger radii; this location therefore marks the horizon. From exercise 1.6 we also know that this isotropic radius $\bar{r}$ corresponds to an areal radius of $R = 2M$, which has a coordinate-independent meaning. We conclude that the horizon of a Schwarzschild black hole is at $R = 2M$, and thus has a proper area of $4\pi (2M)^2$ and a proper circumference of $2\pi (2M)$.

Exercise 1.7 Find the location of the horizon in terms of the isotropic radius r of Exercise 1.5.

Note that the Schwarzschild line element (1.38) becomes singular at $R = 2M$. We have seen already that we can remove this singularity by transforming to a different coordinate system; it is therefore just a *coordinate singularity*, akin to the singular behavior of longitude and latitude (or spherical polar coordinates) at the Earth's north and south poles. For numerical purposes it is important that the metric remains non-singular on the horizon; this is a first hint, therefore, that a coordinate system with properties similar to that in exercise 1.6 might prove very useful in numerical simulations. This is indeed the case, and we will therefore explore the properties of these coordinate systems in more detail in Chapter 5. As a preview, however, we can already construct an embedding diagram for the $\bar{t} = const$ slices of exercise 1.6.

Exercise 1.8 Consider the $\bar{t} = const, \theta = \pi/2$ subspace of the line element (1.47), and equate the resulting two-dimensional line element to (1.41). Match the coefficients in the line elements to find a differential equation for z, and then solve this differential equation to obtain

$$z = (4M(2R - M))^{1/2} + M \ln \frac{(2R - M)^{1/2} - M^{1/2}}{(2R - M)^{1/2} + M^{1/2}}, \qquad (1.52)$$

where we have chosen the positive root and omitted an irrelevant constant of integration.

Hint: Recall, or prove, that $\tanh^{-1} x = (1/2) \ln ((1 + x)/(1 - x))$.

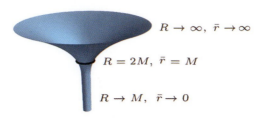

$R \to \infty, \ \bar{r} \to \infty$

$R = 2M, \ \bar{r} = M$

$R \to M, \ \bar{r} \to 0$

Figure 1.7 Embedding diagram for a $\bar{t} = const$ slice of the Schwarzschild line element (1.48), labeled by the radial coordinate $\bar{r}$. As in Fig. 1.6 the black ring marks the event horizon. The shape of this diagram suggests the name "trumpet".

In Fig. 1.7 we show an embedding diagram for the $\bar{t} = const$ slices of the line element (1.48). The shape of this diagram suggests the name *trumpet* for the geometry of these slices. In Section 5 we will discuss in more detail why these geometries have very desirable properties for numerical simulations. We will also return to adopting the coordinates of exercises 1.6 and 1.8 at several places later in this book (see exercises 2.20, 2.27, 3.2, 4.5, and 5.2), using this example of analytical trumpet coordinates to explore new concepts and quantities. For now, however, we will just observe that the geometry of different spatial slices of the Schwarzschild spacetime can be very different – for the $t = const$ slices of the line element (1.38) we obtained a wormhole geometry, while for the $\bar{t} = const$ slices of the line element (1.48) we found a trumpet geometry. Exercise 1.9 provides another remarkable example, for which the geometry of spatial slices is, in fact, flat. Its embedding diagram, accordingly, would simply be a flat surface.

Exercise 1.9 (a) Apply a coordinate transformation of the form

$$dt = d\tilde{t} - f(R)dR \tag{1.53}$$

to the Schwarzschild metric (1.38) and find a function $f(R)$ such that the three-dimensional line element of the $\tilde{t} = const$ slices becomes that of *flat space*, $ds^2 = dR^2 + R^2 d\Omega^2$.

(b) Show that, under the coordinate transformation of part (a), the four-dimensional line element of the Schwarzschild geometry takes the form

$$ds^2 = -\left(1 - \frac{2M}{R}\right) d\tilde{t}^2 + 2\left(\frac{2M}{R}\right)^{1/2} d\tilde{t} dR + dR^2 + R^2 d\Omega^2. \tag{1.54}$$

These coordinates are called *Painlevé–Gullstrand* coordinates.[17]

---

[17] See Painlevé (1921); Gullstrand (1922). Some textbooks refer to these coordinates as "rain-drop" coordinates; see Taylor and Wheeler (2000); Moore (2013). The coordinates of

Evidently, the curvature that a spatial slice "inherits" from the curved spacetime depends on how we "slice" the spacetime – this is a key insight, which we will explore in much more detail in Chapter 2 and beyond.

We conclude this section by noting that the Schwarzschild spacetime can be generalized in a number of different ways. Most importantly, the analytic Kerr metric describes stationary rotating black holes.[18] Without making simplifying assumptions, however, black holes in general cannot be described analytically. For example, the spacetime describing a time-dependent black hole with a non-perturbative distortion is not known analytically, nor is the spacetime describing the inspiral and merger of two black holes in binary orbit. Both examples require the tools of numerical relativity, which we will develop in the following chapters.

1.3.2 Gravitational Waves

Consider a small perturbation of a given "background" metric. Specifically, let us assume that the background metric is the flat Minkowski metric (1.19) in Cartesian coordinates. In that case we can write

$$g_{ab} = \eta_{ab} + h_{ab} \qquad (1.55)$$

and assume that the perturbation h_{ab} is small compared with unity, $|h_{ab}| \ll 1$. Given this metric we can compute the Christoffel symbols, the Riemann tensor, the Ricci tensor, and finally the Einstein tensor. At each step we can drop all terms that are of order higher than linear in h_{ab}, and we can also impose certain coordinate conditions to simplify our expressions. Inserting our result for the Einstein tensor into Einstein's equations (1.37) then yields the linearized field equation appropriate for weak gravitational fields,

$$\Box \bar{h}_{ab} = \eta^{cd} \partial_c \partial_d \bar{h}_{ab} = \left(-\partial_t^2 + D^2 \right) \bar{h}_{ab} = -16\pi T_{ab} \quad \text{(weak-field)}. \qquad (1.56)$$

Here $\bar{h}_{ab} \equiv h_{ab} - \eta_{ab} h/2$ is the "trace-reversed" metric perturbation, where $h = h_c^{\ c} = \eta^{cd} h_{cd}$ is the trace of h_{ab}. Also, ∂_t^2 denotes the second time derivative, D^2 is the spatial flat-space Laplace operator,

---

Exercise 1.6 and the Painlevé–Gullstrand coordinates of this exercise form end-points of an entire one-parameter family of analytical coordinate systems for the Schwarzschild spacetime; see Dennison and Baumgarte (2014).

[18] See Kerr (1963).

and $\Box$ is the D'Alembertian. Together, the two terms $(-\partial_t^2 + D^2)$ form a wave operator, meaning that Einstein's equations reduce to a wave equation for the perturbations h_{ab} in this limit. But where there is a wave equation, there will also be wave-like solutions: we see how gravitational waves emerge very naturally from Einstein's equations.

Even though Einstein recognized the existence of wave-like solutions very early on,[19] uncertainty over the physical reality of gravitational radiation persisted for several decades (caused, in part, by Einstein himself[20]). The issue was clarified in the late 1950s, however,[21] and this triggered the quest to detect gravitational waves from astrophysical sources.

The first observational evidence for gravitational waves was provided by the "Hulse–Taylor" pulsar PSR 1913+16, which has a close binary companion that is also a neutron star.[22] As we will see below, the emission of gravitational radiation leads to a shrinking of the binary orbit. Only a few years after the discovery of the Hulse–Taylor binary, this orbital decay due to gravitational wave emission was not only observed but also shown to be in remarkable agreement with predictions. In 1993, Russell Hulse and Joseph Taylor were awarded a Nobel Prize for this momentous, albeit indirect, confirmation of the existence of gravitational radiation.

Meanwhile, efforts continued to detect gravitational waves directly. Almost exactly one hundred years after Einstein first predicted the existence of gravitational radiation, this quest became reality when the Laser Interferometer Gravitational Wave Observatory (LIGO) detected gravitational waves for the first time on September 14, 2015.[23] Named after the date of detection, the event is called GW150914, and we now are confident that it was emitted by two black holes with masses of about 36 and 29 solar masses that merged about a billion years ago at a distance of about 410 Mpc.[24] This stunning discovery was rewarded with a Nobel Prize to Barry Barish, Kip Thorne, and Rainer Weiss in 2017. Since then, several other gravitational wave signals from merging binary black holes have been detected, as well as signals believed to be

---

[19] See Einstein (1916).
[20] See Kennefick (2005) for an account.
[21] Of particular importance in these developments was the 1957 General Relativity Conference at Chapel Hill; see Bergmann (1957).
[22] See Hulse and Taylor (1975); Taylor and Weisberg (1982).
[23] See Abbott et al. (LIGO Scientific Collaboration and Virgo Collaboration) (2016a); see also Fig. 5.1.
[24] Recall that 1 pc $= 3.0857 \times 10^{13}$ km.

from merging binary neutron stars and possibly even binary black hole–neutron stars. Sources containing neutron stars are particularly exciting, since simultaneous electromagnetic counterpart radiation, when it is observed, provides information on both the location and nature of the merger. This was the case for GW 170817, identified as a binary neutron star merger.[25] The combined gravitational wave and electromagnetic observations represent the "holy grail" of *multimessenger astronomy*.

Now we return to the mathematics of gravitational waves. Since (1.56) is a linear, flat-space wave equation, we can find solutions in terms of the Green function for the flat wave operator. The result for weak fields and slow velocities much less than the speed of light can be written as

$$h_{ij}^{\mathrm{TT}} \simeq \frac{2}{d} \ddot{\mathcal{I}}_{ij}^{\mathrm{TT}}(t - d) \qquad \text{(weak-field, slow-velocity),} \qquad (1.57)$$

where d is the distance from the source and where the double dot denotes a second time derivative. The so-called "reduced quadrupole moment tensor" of the emitting source,

$$\mathcal{I}_{ij} = I_{ij} - \frac{1}{3}\eta_{ij}I, \qquad (1.58)$$

can be computed from the second moment of the mass distribution,

$$I_{ij} = \int d^3x \rho x_i x_j, \qquad (1.59)$$

and its trace $I = I^i_{\ i}$. The "TT" superscript denotes that we must project out the transverse-traceless part of the tensor, using the projection tensor $P_i^{\ j} = \eta_i^{\ j} - n_i n^j$, where $n^i = x^i/d$. We then have

$$\mathcal{I}_{ij}^{\mathrm{TT}} = \left(P_i^{\ j} P_i^{\ j} - \frac{1}{2} P_{ij} P^{kl} \right) \mathcal{I}_{kl}. \qquad (1.60)$$

Exercise 1.10 Consider a *Newtonian* binary consisting of two point masses m_1 and m_2 in circular orbit about their center of mass. Choose coordinates such that the binary system orbits in the xy plane and is aligned with the x-axis at $t = 0$.

---

[25] See Abbott et al. (LIGO Scientific Collaboration and Virgo Collaboration) (2017); see also Fig. 1.9.

(a) Show that the second moment of the mass distribution can be written as

$$I_{ij} = \frac{1}{2}\mu R^2 \begin{pmatrix} 1 + \cos 2\Omega t & \sin 2\Omega t & 0 \\ \sin 2\Omega t & 1 - \cos 2\Omega t & 0 \\ 0 & 0 & 0 \end{pmatrix}, \qquad (1.61)$$

where Ω is the orbital angular velocity, $\mu \equiv m_1 m_2/(m_1 + m_2)$ is the reduced mass, and R is the binary separation.

(b) Show that the reduced quadrupole moment is

$$\mathcal{I}_{ij} = \frac{1}{2}\mu R^2 \begin{pmatrix} 1/3 + \cos 2\Omega t & \sin 2\Omega t & 0 \\ \sin 2\Omega t & 1/3 - \cos 2\Omega t & 0 \\ 0 & 0 & -2/3 \end{pmatrix}. \qquad (1.62)$$

(c) Show that the magnitude of the metric perturbations h_{ij} is approximately

$$h \simeq \frac{4}{d}\frac{\mu M}{R}, \qquad (1.63)$$

where $M = m_1 + m_2$ is the total mass, and where we have used Kepler's third law $\Omega^2 = M/R^3$.

From (1.57) we can also estimate a source's rate of energy loss due to the emission of gravitational radiation,

$$\frac{dE}{dt} = -\frac{1}{5}\langle \dddot{\mathcal{I}}_{ij} \dddot{\mathcal{I}}^{ij} \rangle \qquad \text{(weak-field, slow-velocity)}, \qquad (1.64)$$

where the angle brackets denote a time average. This is the famous *quadrupole formula*, similar to the Larmor formula in electrodynamics, although the latter describes dipole radiation, which is absent in general relativity. A binary system, for example, loses energy in the form of gravitational radiation, which leads to a shrinking of the binary orbit and ultimately to the merger of the two companions. The quadrupole formula (1.64) allows us to make estimates for this inspiral which become precise for weak-field, slow-velocity sources.

Exercise 1.11 (a) Revisit the *Newtonian* binary of exercise 1.10 and show that gravitational wave emission will result in the binary losing energy E at a rate

$$L_{GW} = -\frac{dE}{dt} = \frac{32}{5}\frac{M^3\mu^2}{R^5}. \qquad (1.65)$$

(b) Recall that the equilibrium energy of a binary in circular orbit is given by

$$E_{eq} = -\frac{1}{2}\frac{M\mu}{R}. \qquad (1.66)$$

In response to the energy loss of part (a) the binary separation R will therefore shrink. Assuming that we can consider this inspiral as "adiabatic", i.e. as a sequence of circular orbits along which the radius changes on timescales much longer than the orbital period, we can compute the inspiral rate from

$$\frac{dR}{dt} = \frac{dE_{eq}/dt}{dE_{eq}/dR} = -\frac{L_{GW}}{dE_{eq}/dR} = -\frac{64}{5}\frac{M^2\mu}{R^3}. \qquad (1.67)$$

Verify the last equality.

(c) Integrate equation (1.67) to find the binary separation R as a function of time. Express your result in terms of $T - t$, where T is the time of coalescence, when $R = 0$.

(d) Combine your result from part (c) with Kepler's law $\Omega = (M/R^3)^{1/2}$ to compute the orbital angular velocity Ω as a function of time. To leading order, and for quadrupole emission, the frequency of gravitational waves f_{GW} is twice that of the orbit, $f_{GW} = 2f = \Omega/\pi$. Show that

$$f_{GW} = \frac{5^{3/8}}{8\pi}\frac{1}{\mathcal{M}^{5/8}(T-t)^{3/8}}, \qquad (1.68)$$

where $\mathcal{M} = \mu^{3/5}M^{2/5}$ is the so-called "chirp mass".

Equation (1.68) is a very important result, as it provides the leading-order expression for the gravitational wave frequency of an inspiraling binary as a function of time. Notice that (1.68) depends on a particular combination of the binary's total mass M and its reduced mass μ. namely the *chirp mass* $\mathcal{M} \equiv \mu^{3/5}M^{2/5}$.

A nice way of writing (1.68) is

$$\mathcal{M}f_{GW} = \frac{5^{3/8}}{8\pi}\left(\frac{T-t}{\mathcal{M}}\right)^{-3/8}, \qquad (1.69)$$

since it shows that there is a unique relationship between the non-dimensional quantities $\mathcal{M}f_{GW}$ and $(T-t)/\mathcal{M}$; non-dimensional, that is, in the geometrized units that we have adopted all along. However, equation (1.68) also provides a good opportunity to practice re-inserting the constants G and c,

$$G \simeq 6.673 \times 10^{-8}\ \frac{cm^3}{g\,s^2}, \qquad c \simeq 2.998 \times 10^{10}\ \frac{cm}{s}, \qquad (1.70)$$

in order to obtain an expression in physical units, in this case Hz or s^{-1}. This can be done in many different ways; here is one of them. First observe that we need to eliminate the chirp mass's units of grams. We therefore multiply $\mathcal{M}$ by G and obtain, up to numerical factors, $f \propto (G\mathcal{M})^{-5/8}(T-t)^{-3/8}$. The combination $G\mathcal{M}$ has units cm^3/s^2; dividing

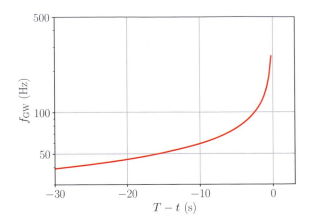

Figure 1.8 The Newtonian quadrupole expression (1.72) for the gravitational wave frequency $f_{\rm GW}$ as a function of time for a binary with chirp mass $\mathcal{M} = 1.188 M_{\odot}$ (see the main text).

this by c^3 therefore gives us a number with units of seconds. Inserting this into (1.68) yields

$$f_{\rm GW} = \frac{5^{3/8}}{8\pi} \left(\frac{c^3}{G\mathcal{M}} \right)^{5/8} \frac{1}{(T-t)^{3/8}}. \qquad (1.71)$$

Finally, we could express the chirp mass in terms of the solar mass $M_{\odot} \simeq 2 \times 10^{33}$ g to obtain

$$
\begin{aligned}
f_{\rm GW} &= \frac{5^{3/8}}{8\pi} \left(\frac{c^3}{G M_{\odot}} \right)^{5/8} \left(\frac{M_{\odot}}{\mathcal{M}} \right)^{5/8} \frac{1}{(T-t)^{3/8}} \\
&\simeq 155.7\,{\rm Hz} \left(\frac{M_{\odot}}{\mathcal{M}} \right)^{5/8} \left(\frac{1\,{\rm s}}{T-t} \right)^{3/8}.
\end{aligned} \qquad (1.72)
$$

In Fig. 1.8 we show an example of this gravitational wave frequency. Notice that the frequency increases with time and also that, for binary masses in the range of solar masses, the frequencies are in the audible range for seconds before the merger. An audio rendering of this signal would sound not unlike a bird's chirp – hence the name "chirp signal".[26]

It is a remarkable development that the relation (1.72) is no longer just a theoretical prediction but has been observed directly. In Fig. 1.9

[26] Examples can be found, for example, at https://www.ligo.caltech.edu/page/gw-sources.

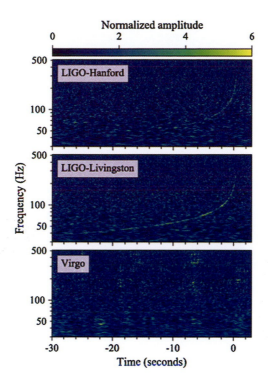

Figure 1.9 A wave-frequency versus time representation of the LIGO and Virgo data for the gravitational wave event GW170817. The signal has been interpreted as originating from a coalescing binary neutron star system with a chirp mass of $\mathcal{M} = 1.188 M_{\odot}$. Note the similarity with the quadrupole result shown in Fig. 1.8. (Figure from Abbott et al. (LIGO Scientific Collaboration and Virgo Collaboration) (2017).)

we show Fig. 1 of Abbott et al. (LIGO Scientific Collaboration and Virgo Collaboration) (2017), which announced the first ever direct detection of gravitational waves from a merging binary neutron star system. The signal, now called GW170817, clearly shows a "chirp-signal" very similar to our estimate (1.72) shown in Fig. 1.8. This is truly amazing!

We can qualitatively describe the gravitational waveform from inspiraling binaries as follows (see also Fig. 1.10): Emitting gravitational radiation extracts both energy and angular momentum from the binary. The loss of energy leads to a shrinking of the binary separation, as we estimated in (1.67). According to Kepler's law, the binary's orbital frequency, and hence the frequency of the emitted gravitational radiation,

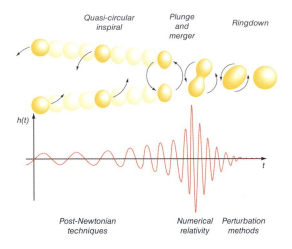

Figure 1.10 The different phases of compact binary inspiral and coalescence. The gravitational wave amplitude $h(t)$ is shown schematically, and the analysis technique is identified for each phase. (Figure from Baumgarte and Shapiro (2010).)

increase as the binary separation shrinks. As we estimated in (1.63), the amplitude of the emitted gravitational radiation also increases with decreasing binary separation. During the inspiral, therefore, both the amplitude and frequency of the gravitational wave signal increase. From the quadrupole formula (1.64) we can see also that the rate of energy emission continuously increases, meaning that the inspiral proceeds increasingly fast as the binary separation decreases – see (1.67).

At some point, of course, the above description breaks down. We can crudely estimate this point by checking when the inspiral no longer remains adiabatic, i.e. at what point the orbital radius no longer shrinks on timescales much longer than the orbital period, as we assumed in part (b) of exercise 1.11. We can find the orbital decay timescale τ_{GW} from

$$\tau_{\mathrm{GW}} = \left| \frac{R}{dR/dt} \right| = \frac{5}{64} \frac{R^4}{M^2 \mu}, \qquad (1.73)$$

where we have used (1.67) in the last equality. Our approximation breaks down when this orbital decay timescale becomes equal to the orbital period $P_{\mathrm{orb}} = 2/f_{\mathrm{GW}}$, which happens at a critical frequency $f_{\mathrm{GW}}^{\mathrm{crit}}$ that depends on the binary's chirp mass $\mathcal{M}$. We can think of this critical frequency as the approximate frequency at which the slow inspiral transitions to a dynamical plunge and merger, as shown in Fig. 1.10. The amplitude h of the emitted gravitational wave signal will also assume its maximum value close to this transition, when the objects merge.

Exercise 1.12 (a) Express τ_{GW} in (1.73) in terms of the gravitational wave frequency f_{GW} and equate your result to P_{orb} to show that the approximation of an adiabatic inspiral breaks down for a critical frequency f_{GW}^{crit} given by

$$f_{GW}^{crit} = \frac{5^{3/5}}{2^{21/5}\pi^{8/5}} \frac{1}{\mathcal{M}}. \tag{1.74}$$

(b) Use Kepler's law to show that the approximate binary separation R corresponding to the critical orbital frequency $\Omega^{crit} = \pi f_{GW}^{crit}$ is given by

$$R^{crit} = \frac{2^{14/5}\pi^{2/5}}{5^{2/5}} M^{3/5}\mu^{2/5} \simeq 5.8 M^{3/5}\mu^{2/5}. \tag{1.75}$$

(c) Use (1.63) to show that a crude estimate for the maximum gravitational wave amplitude, as measured at a distance d, is

$$h^{crit} \simeq \frac{5^{2/5}}{2^{4/5}\pi^{2/5}} \frac{\mathcal{M}}{d} \simeq 0.69 \frac{\mathcal{M}}{d}. \tag{1.76}$$

(d) Show that, in physical units, the critical frequency (1.74) becomes

$$f_{GW}^{crit} \simeq 4.9\,\text{kHz} \left(\frac{M_\odot}{\mathcal{M}}\right), \tag{1.77}$$

with wave amplitude (1.76) given by

$$h^{crit} \simeq 3.3 \times 10^{-20} \left(\frac{\mathcal{M}}{M_\odot}\right)\left(\frac{\text{Mpc}}{d}\right). \tag{1.78}$$

Evidently, our adiabatic approximations are valid only for frequencies well below the critical frequency (1.77).

The LIGO and Virgo gravitational wave detectors are most sensitive in the range of tens to hundreds of hertz. For binary neutron stars, which have chirp masses on the order of a solar mass, these frequencies are reasonably well below the critical frequency (1.77). For these binaries, current gravitational wave detectors can observe the late inspiral but not the merger itself. In the range where these detectors can measure the signal, our adiabatic approximations are therefore reasonably accurate, which explains the similarity between the curves in Figs. 1.8 and 1.9. Since our Newtonian quadrupole result (1.72) dominates the signal in this regime, at least for sufficiently large binary separations and hence sufficiently small frequencies, and since this contribution depends on the chirp mass only, it is this particular combination of the binary's masses that can be determined most accurately. For the signal GW170817, for example, it was found that $\mathcal{M} = 1.188^{+0.004}_{-0.002} M_\odot$. Higher-order corrections to the signal can be computed from non-linear post-Newtonian expansions, which improve the accuracy of the above estimates for smaller binary separations and hence higher frequencies; these higher-order corrections then provide information about the *individual* masses of the binary companions (as well as, at higher orders yet, their spin and internal structure). Post-Newtonian approximations also break down once the neutron stars coalesce and merge, and that is where we will need the tools of numerical relativity to model these processes.

Binary black holes, however, are expected to have larger chirp masses. For binaries consisting of stellar-mass black holes, for example, with masses around 10 to 50 $M_\odot$, we expect the chirp mass to be on the order of tens of solar masses as well. The critical frequency (1.77) then drops to a few hundred Hz or so, implying that gravitational wave observations made with the currently available detectors will not remain for very long in the range in which our adiabatic estimates are accurate. Instead, the merger of such binaries is expected to occur at frequencies within the detectors' most sensitive range – which is about what happened for the now-famous, first ever detected signal GW150914 (see Fig. 5.1).[27] Clearly, the approximations in this section cannot be used to predict or analyze these signals accurately, and even post-Newtonian approximations are not sufficient to model the dynamical merger. In order to model these processes we therefore rely on numerical relativity – which we will develop in the following chapters.

---

[27] The planned space-based instrument, the Laser Interferometer Space Antenna (LISA), is designed to detect gravitational waves in the range 10^{-4}–10^{-1} Hz, and is therefore sensitive to signals from merging intermediate and supermassive black-hole binaries in the range of about 10^3–$10^6 M_\odot$.

2

Foliations of Spacetime: Constraint and Evolution Equations

In Chapter 1 we motivated Einstein's equations and saw how elegantly they relate the curvature of spacetime to its mass–energy content. From a numerical perspective, however, this elegance is somewhat of a hindrance. In Einstein's equations, all quantities are spacetime quantities, and space and time are neatly placed on the same footing in the fundamental field equation (1.37). In numerical work, however, typically we would like to solve the Cauchy *or* initial-value *problem, starting with a certain state of the fields and matter at some initial time and then following how the fields and matter evolve as time advances. This requires splitting spacetime into space and time, i.e. introducing a so-called* foliation *of spacetime, or a* 3+1 *split. Leaning heavily on analogies with scalar fields and electrodynamics we will see that this split results in* constraint *and* evolution *equations: the former constrain the fields at each instant of time, including the initial time, while the latter determine the evolution of the fields as time advances.*

2.1 Scalar Fields

We will start our discussion with a simple massless scalar field in flat spacetime, in order to illustrate a few points and to provide a reference point for our discussion later on. The field equation for a massless scalar field ψ is the wave equation[1]

$$\Box \psi \equiv \nabla^a \nabla_a \psi = 4\pi \rho, \tag{2.1}$$

---

[1] A scalar field for a particle with *finite* rest mass m satisfies the *Klein–Gordon* equation $\Box \psi - m^2 \psi = 4\pi \rho$. Here we have assumed *natural units*, with $c = 1$ and $\hbar = 1$, which are commonly used in the context of field theory. In Appendix C.2 we show how these equations of motion result from the vanishing of the divergence of the stress–energy tensor for a scalar field.

where we have allowed for a non-zero matter source term with total mass–energy density ρ on the right-hand side. In a flat Minkowski spacetime we can replace g_{ab} with η_{ab} and then expand the wave operator on the left-hand side as follows:

$$\Box\psi = g^{ab}\nabla_a\nabla_b\psi = \eta^{ab}\nabla_a\nabla_b\psi = \left(-\partial_t^2 + D^2\right)\psi. \qquad (2.2)$$

This equation deserves a few words of explanation. First, we should remind the reader that g^{ab} is the inverse of the metric (see the discussion below equation 1.29). In the third equality we have split the sum over the spacetime indices a and b into a term for the time index t ($a,b = 0$), and the remaining spatial terms ($a,b = i,j$). The latter terms result in the spatial Laplace operator, or Laplacian, which we denote as D^2. We will generalize the above in the following sections, but for now we can evaluate this Laplace operator in the same way as in flat space. In Cartesian coordinates all derivatives reduce to partial derivatives, and for curvilinear coordinates we can look up the Laplace operator in electrodynamics textbooks.[2] In spherical polar coordinates, for example, the Laplace operator becomes the familiar expression (A.49), derived in Exercise A.9.

Note, by the way, that (2.1) reduces to the Newtonian field equation (1.13) if we can neglect the time derivatives. We also recognize that the field equation (2.1) for the scalar field again involves second derivatives, exactly as in the Newtonian field (Poisson) equation (1.13) for Φ and as in Einstein's field equations (1.37) for the spacetime metric g_{ab}.

Equation (2.1) is a second-order partial differential equation for ψ. Mathematicians classify this equation as a symmetric hyperbolic equation and assure us of its very desirable properties: the initial value or Cauchy problem is well-posed, meaning that solutions exist, are unique, and grow at most exponentially.

From a numerical perspective, it is easier to integrate equations that are first-order in time. We could therefore define a new variable

$$\kappa \equiv -\partial_t\psi, \qquad (2.3)$$

where inclusion of the minus sign will make sense later on, and rewrite equation (2.1) as the pair of first-order equations[3]

---

[2] See, e.g., Jackson (1999); Griffiths (2013); Zangwill (2013).
[3] In fact, many authors would introduce new variables for the first spatial derivatives as well, thereby making the system first order in both space and time.

$$\partial_t \psi = -\kappa, \tag{2.4}$$

$$\partial_t \kappa = -D^2 \psi + 4\pi \rho. \tag{2.5}$$

The color coding is intended to make comparisons with electrodynamics and general relativity easier, as we will see shortly. We apologize to readers who cannot distinguish these colors, and emphasize that they are not crucial to follow the discussion. We identify the different colors assigned to the variables in Box 2.2 below, which can be checked whenever reference is made to variables of a specific color.

The key point is this: it's pretty simple to construct numerical solutions for the scalar field equation. For starters we note that there are no *constraint equations*. This means that we can choose *initial data* for ψ and κ freely; these initial data, given as functions of space only, then describe ψ and κ at some initial instant of time. Equations (2.4) and (2.5) are the *evolution equations* that determine how these variables evolve as time advances. It is not very hard to write a computer program that solves the evolution equations (2.4) and (2.5) numerically; given the fields at an "old" time t, such a program will recursively use the evolution equations to compute the fields at a "new" time $t + dt$, then again at $t + 2dt$, etc. We will discuss some suitable numerical strategies in Appendix B.3, and will provide an example code in B.3.2 – albeit for Maxwell's equations rather than a scalar field. For our purposes here, however, the details of the numerics are not important.

One could argue that equations (2.4) and (2.5) represent a 3+1 *split* of the wave equation (2.2), since we started with an operator involving both time and space derivatives on the left-hand side of that equation and ended up with a pair of equations in which we have explicitly separated the three space derivatives from the one time derivative – hence "3+1". In this example there was so little to do, though, that one easily misses the point. For a starter, there were no mixed time and space second derivatives. More importantly, we only had to deal with the derivatives in the equation, but, since the variable here was a scalar, we did not have to work to separate any time components from the space components. However, what do we need to do if the variables are tensors, so that we may also want to split the tensors into separate time and space components? As a warm-up exercise we will consider electrodynamics in the next section, before discussing general relativity in Section 2.3. At each step along the way, and more systematically in Section 2.4, we will make comparisons with the scalar field equations (see, in particular, Boxes 2.1 and 2.3 for a direct comparison between the different sets of equations); we will find remarkable similarities with the equations (2.4) and (2.5) above but also important differences. The similarities will help guide our intuitive understanding of the equations, while the differences will point to some issues that arise in numerical relativity, which we will discuss in Chapters 3 and 4.

2.2 Electrodynamics

2.2.1 Maxwell's Equations

Maxwell's equations for the electric field $\mathbf{E}$ and the magnetic field $\mathbf{B}$ are

$$\begin{aligned}\mathbf{D} \cdot \mathbf{E} &= 4\pi\rho, & \partial_t \mathbf{E} &= \mathbf{D} \times \mathbf{B} - 4\pi\mathbf{j}, \\ \mathbf{D} \cdot \mathbf{B} &= 0, & \partial_t \mathbf{B} &= -\mathbf{D} \times \mathbf{E},\end{aligned} \qquad (2.6)$$

where we have used vector notation: $\mathbf{D}$ is the spatial gradient operator (rather than the more common symbol ∇, which we are reserving for four dimensions), ρ is the charge density, and $\mathbf{j}$ is the charge current density. The top left equation in (2.6) is also known as Gauss's law, the top right as Ampère's law (with Maxwell's correction), and the bottom right as Faraday's law. The bottom left equation is the "no magnetic monopoles" equation and tells us about an important property of the magnetic field: it is always divergence-free, meaning that we may write it as the curl of a vector potential $\mathbf{A}$, i.e.

$$\mathbf{B} = \mathbf{D} \times \mathbf{A}. \qquad (2.7)$$

The divergence of $\mathbf{B}$ then vanishes identically,

$$\mathbf{D} \cdot \mathbf{B} = \mathbf{D} \cdot (\mathbf{D} \times \mathbf{A}) = 0, \qquad (2.8)$$

as required. The curl of the magnetic field in Ampère's law now becomes

$$\mathbf{D} \times \mathbf{B} = \mathbf{D} \times (\mathbf{D} \times \mathbf{A}) = -D^2\mathbf{A} + \mathbf{D}(\mathbf{D} \cdot \mathbf{A}). \qquad (2.9)$$

Finally, Faraday's law becomes

$$\partial_t(\mathbf{D} \times \mathbf{A}) = \mathbf{D} \times (\partial_t \mathbf{A}) = -\mathbf{D} \times \mathbf{E}. \qquad (2.10)$$

Since the curl of a gradient vanishes identically, this implies that

$$\partial_t \mathbf{A} = -\mathbf{E} - \mathbf{D}\phi, \qquad (2.11)$$

where ϕ is an arbitrary gauge variable. Collecting our results, we see that Maxwell's equations (2.6) result in the two equations

$$\partial_t \mathbf{A} = -\mathbf{E} - \mathbf{D}\phi, \tag{2.12}$$

$$\partial_t \mathbf{E} = -D^2 \mathbf{A} + \mathbf{D}(\mathbf{D} \cdot \mathbf{A}) - 4\pi \mathbf{j}, \tag{2.13}$$

or, returning to index notation,

$$\partial_t A_i = -E_i - D_i \phi, \tag{2.14}$$

$$\partial_t E_i = -D^2 A_i + D_i D^j A_j - 4\pi j_i. \tag{2.15}$$

We notice immediately the similarity with equations (2.4) and (2.5) for the scalar wave: in both cases the first equation relates the time derivative of the "blue" variable to the "red" variable, and the second equation relates the time derivative of the red variable to the Laplace operator acting on the blue variable, plus a "green" matter term. In fact, we can now appreciate the inclusion of the negative sign in the definition (2.3), which makes even the signs consistent. We also notice important differences, in addition to the fact that in equations (2.4) and (2.5) the dynamical variables were scalars while in (2.14) and (2.15) they are vectors: in the electromagnetic case we encounter a new gauge variable in (2.14), highlighted in a gold color, and (2.15) includes not only the Laplace operator of the blue variable but also the gradient of the divergence. The latter will play an important role later on. Without this term, equations (2.14) and (2.15) could be combined to form a simple wave equation for the components E_i, but these new terms spoil this property.

As another difference, we also still have to satisfy Gauss's law

$$D_i E^i = 4\pi \rho. \tag{2.16}$$

Unlike (2.14) and (2.15), this equation does not contain any time derivatives of the variables; instead, the electric field E_i has to satisfy this equation at all times. We refer to this equation as a *constraint equation*. In particular, any initial data that we choose for E_i have to satisfy the constraint equation (2.16) – we can no longer choose the initial data arbitrarily. Equations (2.14) and (2.15) do contain time derivatives of the fields, however; they therefore determine how the fields evolve forward in time, and so they are called *evolution equations*. Exercise 2.1 demonstrates that the constraint equation is *preserved* by the evolution equations, meaning that if the constraint equation is satisfied initially, it will be satisfied at all times – at least analytically.

Exercise 2.1 Define a constraint violation function $\mathcal{C} \equiv D_i E^i - 4\pi\rho$ and assume that charge is conserved, so that the continuity equation $\partial_t \rho = -D_i j^i$ holds. Then show that the time derivative of the constraint violation function vanishes, $\partial_t \mathcal{C} = 0$.

The above decomposition already illustrates the general procedure that is used to solve Einstein's equations numerically. As we will discuss in Section 2.3, Einstein's equations can also be split into a set of constraint equations and a set of evolution equations, where the latter explicitly involve time derivatives of the fields while the former do not. In fact, the equations will have a remarkable similarity to the set of equations given above. We again have to solve the constraint equations first in order to construct valid initial data. These initial data will describe an instantaneous state of the gravitational fields at some instant of time – they might, for example, describe a snapshot of two black holes that are in the process of merging. Then we can solve the evolution equations to compute how the fields evolve in time – this would result in a "movie" of the subsequent coalescence together with the emission of a gravitational wave signal.

The observant reader will notice, however, that the above example has not yet accomplished one goal: namely, to provide an example of how we can split a spacetime tensor into space and time parts. Our example does not do that, because we started with Maxwell's equations already expressed in terms of purely spatial, three-dimensional, electric and magnetic field vectors. Fortunately, it is easy to fix that problem – we could start with Maxwell's equations expressed in terms of the spacetime, four-dimensional Faraday tensor instead of these three-dimensional electric and magnetic fields.

2.2.2 The Faraday Tensor

Maxwell's equations show that electric and magnetic fields are intimately related. The Lorentz transformations for electric and magnetic fields also show that one observer's electric field may manifest itself as a magnetic field to another observer, and vice versa. In particular, this shows that electric and magnetic fields do *not* transform like components of vectors. What kind of a relativistic object, then, do the fields form? Just counting the number of independent components gives us a hint.

Evidently, there are six independent components in the combined electric and magnetic fields, more than a four-vector (a rank-1 tensor in four dimensions) could accommodate. A rank-2 tensor in four dimensions has $4 \times 4 = 16$ independent components – evidently that is too many. The metric tensor, for example, is a *symmetric* rank-2 tensor, with $g_{ab} = g_{ba}$. In four dimensions, such a tensor has 4 (diagonal) + 6 (off-diagonal) = 10 independent components – still too many. An *antisymmetric* rank-2 tensor A_{ab} has the property $A_{ab} = -A_{ba}$; in particular, this means that the diagonal components A_{aa} have to vanish. In four dimensions, such a tensor therefore has only six (off-diagonal) independent components – exactly the number that we need to accommodate the electric and magnetic fields. That is a smoking gun. In fact, the covariant object describing electric and magnetic fields is indeed an antisymmetric rank-2 tensor, called the *Faraday* tensor.

An observer using Cartesian coordinates (in a local Lorentz frame) would identify the components of the Faraday tensor F^{ab} as follows:

$$F^{ab} = \begin{pmatrix} 0 & E^x & E^y & E^z \\ -E^x & 0 & B^z & -B^y \\ -E^y & -B^z & 0 & B^x \\ -E^z & B^y & -B^x & 0 \end{pmatrix}. \tag{2.17}$$

Since the Faraday tensor is antisymmetric, we now have to be careful with the identification of the components. By convention, we will let the first index in F^{ab} refer to the row, and the second to the column. For example, we identify $F^{tx} = E^x = -F^{xt}$.

We can also introduce the electric and magnetic field four-vectors, E^a and B^a, respectively, where for an observer with a four-velocity u^a we require $u_a E^a = 0$ and $u_a B^a = 0$. For such an observer the Faraday tensor can also be written as

$$F^{ab} = u^a E^b - u^b E^a + u_d \epsilon^{dabc} B_c, \tag{2.18}$$

where ϵ^{abcd} is the completely antisymmetric *Levi–Civita* tensor. In a local Lorentz frame, it equals $+1$ for all even permutations $abcd$ of $txyz$, -1 for all odd permutations, and 0 for all other combinations of indices.

Exercise 2.2 (a) Demonstrate that (2.18) reduces to (2.17) in the observer's local Lorentz frame, where $u^a = (1,0,0,0)$.

(b) Show that an observer with (arbitrary) four-velocity u^a will measure the electric field to be $E^a = F^{ab} u_b$ and the magnetic field to be $B^a = \epsilon^{abcd} u_b F_{dc}/2$.

Hint: For the latter, use $\epsilon^{abcd} \epsilon_{cdef} = -2\delta^{ab}_{ef} \equiv -2(\delta^a_e \delta^b_f - \delta^a_f \delta^b_e)$.

A tensor has well-defined properties under coordinate transformations (see Appendix A). For example, the Faraday tensor components in a new, primed, coordinate system can be found by applying (A.26). To find the transformation of the electric and magnetic fields under a Lorentz transformation, we identify the primed fields with the corresponding components of the primed Faraday tensor – for example $E^{x'} = F^{t'x'}$. Exercise A.4 demonstrates that electric and magnetic fields "mix" under a Lorentz transformation – what appears like a pure electric field in one reference frame is mixed with a magnetic field in another, boosted, reference frame, and vice versa.

In terms of the Faraday tensor, Maxwell's equations can be cast in the compact form

$$\nabla_b F^{ab} = 4\pi j^a \tag{2.19}$$

and

$$\nabla_{[a} F_{bc]} = 0, \tag{2.20}$$

where the current four-vector j^a has components

$$j^a = (\rho, j^i) \tag{2.21}$$

and where the bracket [ ] denotes the antisymmetrization operator.[4] Equations (2.19) and (2.20) are equivalent to equations (2.6). In Appendix C.1 we show how the equation of motion (2.19) results from the vanishing of the divergence of the electromagnetic stress–energy tensor.

Note that the divergence of the left-hand side of equation (2.19) must vanish identically,

$$\nabla_a \nabla_b F^{ab} = 0. \tag{2.22}$$

To see this, assume a Minkowski spacetime for simplicity and adopt Cartesian coordinates, so that the covariant derivatives reduce to partial derivatives $\partial_a \partial_b F^{ab}$. The partial derivatives then commute, meaning that $\partial_a \partial_b$ is symmetric, while F^{ab} is antisymmetric, so that the contraction $\partial_a \partial_b F^{ab}$ must vanish (see part (b) of Exercise A.2).

The identity (2.22) plays the same role in electrodynamics as the Bianchi identities play in general relativity; we mentioned these in Section 1.2.3 and they will reappear in equation (2.97) below.[5] For our

---

[4] See equation (A.16) and the remarks following it.
[5] Different identifications between electrodynamics and general relativity are possible, and which one is most useful depends on the context. In the context of asymptotic structure and radiation, for example, it is useful to identify the electric and magnetic fields with the electric and

purposes we will therefore refer to (2.22) as the "electromagnetic Bianchi identity". As we will see, in both theories these identities have two important properties: they imply a conservation law, and also tell us something profound about constraints and gauge freedom. For the first property, we observe that, if the divergence of the left-hand side of (2.19) vanishes then so must the divergence of the right-hand side,

$$\nabla_a j^a = 0. \tag{2.23}$$

Inserting the components (2.21) we can expand this as

$$\partial_t \rho = -D_i j^i, \tag{2.24}$$

which we recognize as the continuity equation. Evidently charge conservation, which is encoded in the continuity equation, follows directly from (2.19) together with the electromagnetic Bianchi identity (2.22).

Previously we observed that the divergence of the magnetic field B^i vanishes, and we therefore wrote B^i as the curl of a vector potential A^i (see equation 2.7). Similarly, equation (2.20) allows us to write the Faraday tensor in terms of a four-dimensional vector potential A^a,

$$F^{ab} = \nabla^a A^b - \nabla^b A^a. \tag{2.25}$$

Equation (2.20) is then satisfied identically, and the remaining Maxwell equation (2.19) takes the form

$$\nabla_b \nabla^a A^b - \nabla_b \nabla^b A^a = 4\pi j^a. \tag{2.26}$$

In fact, this form fits in well with our outline of Newton's and Einstein's gravity in Chapter 1. In the parlance of our discussion there, the four-dimensional vector potential A^a is now the fundamental quantity of our theory – the analogue of the Newtonian potential Φ in Newtonian gravity and of the spacetime metric g_{ab} in general relativity. The vector potential A^a is a rank-1 tensor, conveniently "halfway" between the rank-0 Newtonian potential Φ (or the scalar field ψ in the wave equation 2.1) and the rank-2 metric g_{ab} in general relativity. Electromagnetic forces on charges result from the electric and magnetic fields in the Lorentz force and hence from derivatives of A^a. Finally, the field equations (2.26) contain the trace of second derivatives of A^a, in complete analogy to the Newtonian Poisson equation (1.15) (or its generalization, the wave equation 2.1) and Einstein's equation (1.37). In fact, we can

---

magnetic parts of the Weyl tensor, in which case equation (2.20) rather than (2.22) plays the role of the Bianchi identities (see, e.g., Newman and Penrose (1968)).

think of the left-hand side of (2.26) as the electromagnetic equivalent of
the Einstein tensor. Equation (2.26) therefore provides a perfect "warm-
up" problem for 3+1 decompositions.

Now we will get to the second consequence of the electromagnetic
Bianchi identity (2.22). Naively, we might expect that equation (2.26)
gives us four wave-type equations for the four components of the four-
vector A^a, suggesting that we are all set to proceed. On second thoughts,
however, this cannot be correct. Because of (2.22), the divergence
of the left-hand side of (2.26) vanishes identically, so that the four
components of the equation cannot all be independent – one equation
must be redundant in relation to the others. This redundancy results
in a constraint equation. By the same token, we then have only three
independent equations for the four components of A^a, meaning that
one component remains undetermined by equations (2.26) – this is
the origin of the gauge freedom in electromagnetism. We therefore see
that both the existence of a constraint equation and gauge freedom are
consequences of the identity (2.22), and we also see that there must be
as many degrees of gauge freedom as there are constraint equations.

To see how this works in detail we start with a 3+1 decomposition
of the tensors appearing in (2.26). In (2.21) we have already split
the four-dimensional current density vector j^a into the charge density
ρ as its time component and the spatial current density j^i as its
spatial components. Similarly, we can write the four-dimensional vector
potential A^a as

$$A^a = (\phi, A^i). \tag{2.27}$$

This completes the decomposition of the tensors, and we next decompose the equations.

We first consider the *time* component of equation (2.26). Choosing
$a = T$ in equation (2.26) results in

$$\nabla_b \nabla^T A^b - \nabla_b \nabla^b A^T = \nabla_j \nabla^T A^j - \nabla_j \nabla^j \phi = 4\pi j^T = 4\pi\rho. \tag{2.28}$$

Here, we have intentionally denoted the time with a capital T in order
to single out this time coordinate in what follows. Note also that we
were able to restrict the summation to the spatial indices $b = j$,
because for $b = T$ the two first terms cancel each other out. This is
consistent with our argument above, that we should not expect to obtain
four independent wave equations. In special relativity we may identify

$\nabla_j = D_j$ and $\nabla_T = \partial_T$. We also have $\nabla^T = \eta^{Ta}\nabla_a = -\partial_T$, and therefore

$$D_j(-\partial_T A^j - D^j\phi) = 4\pi\rho. \qquad (2.29)$$

We now make the identification

$$E^i = F^{Ti} = -\partial_T A^i - D^i\phi, \qquad (2.30)$$

which is totally consistent with (2.14), and obtain

$$D_i E^i = 4\pi\rho, \qquad (2.31)$$

i.e. the constraint equation (2.16).

We next consider the *spatial* components $a = i$ in (2.26) and find

$$\nabla_T(\nabla^i A^T - \nabla^T A^i) = -\nabla_j\nabla^i A^j + \nabla_j\nabla^j A^i + 4\pi j^i, \qquad (2.32)$$

where we have grouped the two terms with $b = T$ on the left-hand side and the two terms with $b = j$ on the right-hand side. We will restrict our analysis to special relativity again, identify A^T with ϕ, insert (2.30) on the left-hand side, and obtain

$$\partial_T E_i = -D_j D^j A_i + D_i D^j A_j - 4\pi j_i, \qquad (2.33)$$

where we have used the fact that, in flat spacetime, we may commute the second derivatives, so that $D_j D_i A^j = D_i D_j A^j = D_i D^j A_j$. Not surprisingly, equation (2.33) is identical to equation (2.15).

While, so far, we have only rederived familiar equations, we can already make the following observations: in the form (2.26), Maxwell's equations are written in terms of a four-dimensional equation for the four-dimensional tensor object A^a, our "fundamental quantity". We performed a 3+1 split by first breaking the tensors into their space and time parts; in particular, we observed that the time component of A^a plays the role of a gauge variable, while each spatial component corresponds to one of the spatial dynamical variables. The time derivatives of these three spatial components A^i are related to the other dynamical variables, the components of the electric field. We then saw that the time component of our field equation (2.26) resulted in the constraint equation, while the spatial components resulted in the evolution equation. The reader will not be shocked to hear that we will find a very similar structure in general relativity. Using that analogy we could jump to the treatment of general relativity given in Section 2.3

right now. Doing that, however, we would miss out on the opportunity to explore now several concepts that will be even more important in general relativity.

2.2.3 "Spacetime Kinematics": the 3+1 Split of Spacetime

In Section 2.2.2 we introduced the notion of a 3+1 split of four-dimensional tensors and equations, but we did so by selecting specific components T and j in the tensors and equations. Consequently, we have not yet taken advantage of the full coordinate freedom at our disposal. Essentially, we assumed that the spatial slices of our 3+1 decomposition have to line up with a given coordinate time T. We will now relax this assumption and will allow for more general slices and coordinates. While this general coordinate freedom is usually not needed for solving Maxwell's equations in flat spacetime, it has in fact proven to be crucial for solving the initial-value or Cauchy problem for many problems in numerical relativity. Many mathematical operators and geometric objects that we encounter here will make a reappearance when we introduce a similar 3+1 decomposition of general relativity in Section 2.3.

Let us assume that we are given a function $t = t(x^a)$ that depends, in some way, on our original coordinates. We have suggestively called this function t, because shortly this will indeed become our new coordinate time, different in general from the quantity that we called T in the previous section.[6] It might be helpful to think of the level surfaces of constant t as three-dimensional "contour surfaces" that stack up to cover our four-dimensional spacetime.[7] We refer to such surfaces as *hypersurfaces* of the spacetime. We will assume that these level surfaces of constant t are spacelike, and will refer to them as *time slices*. As always, the gradient of t, i.e. $\nabla_a t = \partial_a t$, is perpendicular to the level surfaces and points in the direction of fastest change in t. We then introduce the *normal* n_a to the slices as proportional to the gradient,

$$n_a = -\alpha \nabla_a t, \tag{2.34}$$

---

[6] It is not necessary to make t the new coordinate time, and for some applications it may even be advantageous not to make this choice (see also footnote 14 below). We will adopt this assumption nevertheless, since it simplifies our treatment (see also Section 2.7 in Baumgarte and Shapiro (2010) for a discussion).

[7] Our level surfaces may not cover the entire spacetime globally, if, for example, we encounter a coordinate or spacetime singularity.

with unit magnitude

$$n_a n^a = g^{ab} n_a n_b = -1. \tag{2.35}$$

By our assumption that the slices are spatial, the normal four-vector must be timelike and we therefore normalize it to negative unity in (2.35). Also, we will see shortly that the normalization factor α in (2.34) has a very tangible meaning, as it measures the ratio between the advance of proper time and the coordinate time (see equation 2.39 below). In general, α is a function of the coordinates, $\alpha = \alpha(x^a)$, and thus may vary in space and time. Finally, the choice of the negative sign in (2.34) makes the "upstairs" contravariant time component of n^a, which we will encounter in (2.40) below, positive, meaning that this choice makes the normal vector future-oriented.

Note that the normalization (2.35) allows us to interpret the normal vector as a four-velocity, i.e. $u^a = n^a$. We therefore refer to a *normal observer* as an observer whose four-velocity is n^a, and who hence moves normal to our spacelike level surface. More formally we say that the normal observer's worldline is generated by the normal vector n^a.

In our new coordinate system, where t serves as our new time coordinate x^0, the gradient $\nabla_a t = \partial_a t = \partial t / \partial x^a$ becomes particularly simple to evaluate. The only non-zero component is then the component for which $a = t$, i.e. the time component $\partial_t t = 1$, and the normal (2.34) reduces to

$$n_a = (-\alpha, 0, 0, 0). \tag{2.36}$$

It may be helpful to explore the above concepts by working through the concrete example given in exercise 2.3 and shown in Fig. 2.1; we will return to this example several times in this chapter (see exercises 2.4, 2.6, 2.8, 2.21, 2.24, and 2.25 below).

Exercise 2.3   Consider a time coordinate t that is related to the time T by

$$t = T - h(r). \tag{2.37}$$

Here $h(r)$, assumed to depend on the radius $r = (x^2 + y^2 + z^2)^{1/2}$ only, is sometimes called a *height function*; it describes how far the $t = const$ slice lies above the $T = const$ slice, as illustrated in Fig. 2.1. Refer to the coordinates $x^{a'} = (T, r', \theta', \varphi')$ as primed and the coordinates $x^a = (t, r, \theta, \varphi)$ as unprimed, where we set $(r, \theta, \varphi) = (r', \theta', \varphi')$. In the primed frame the metric is given by the Minkowski metric, $g_{a'b'} = \eta_{a'b'}$.

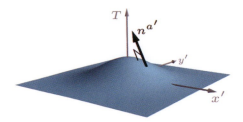

Figure 2.1 Illustration of a spatial hypersurface of constant t. The surfaces of constant time T of Section 2.2.2 would appear as flat horizontal surfaces, but now we consider surfaces of constant t, where $t = T - h(r)$, as in exercise 2.3. Here the height function h is given by $h(r) = A \exp(-(r/\sigma)^2)$, where A and σ are positive constants and $r = r' = \left((x')^2 + (y')^2 + (z')^2\right)^{1/2}$. We show a normal vector $n^{a'} = (n^T, n^{i'})$. The normal vector is tilted with respect to the T-axis since $n^T > 0$ while $n^{r'} < 0$, as shown in exercise 2.3 (see also exercise 1.3).

(a) Evaluate (2.34) to find the components $n_{a'}$ and $n^{a'}$ of the normal vector – normal to the $t = const$ surfaces, that is – in the primed frame.

(b) Normalize the normal vector according to (2.35) to find α.

(c) Now transform the normal vector to the unprimed frame using $n_a = (\partial x^{b'}/\partial x^a) n_{b'}$ (see equation A.29 in Appendix A). If all goes well you will recover (2.36).

Recall that we can always compute the difference in the values of a function at two nearby points by taking the dot product between the gradient of the function and the vector pointing from one point to the other, as in (A.30). Specifically, consider two events separated by an infinitesimal proper time $d\tau$ along a normal observer's worldline, so that the vector pointing from one event to the other is the four-vector $n^a d\tau$. Taking the function to be the coordinate time t we can then measure its advance dt between the two events from

$$dt = \left(\frac{\partial t}{\partial x^a}\right)(n^a d\tau) = (\nabla_a t)(n^a d\tau) = -\frac{1}{\alpha}n_a n^a d\tau = \frac{d\tau}{\alpha} \quad (2.38)$$

or

$$d\tau = \alpha dt. \quad (2.39)$$

Evidently, the function α measures how much proper time elapses, according to a normal observer, as the coordinate time advances by dt – see Fig. 2.2 for an illustration. We therefore call α the *lapse function*.

We have identified our spatial slices – these give the normal n_a an invariant meaning – and so we have used up our freedom to

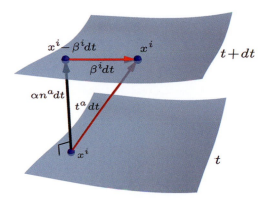

Figure 2.2 Illustration of the lapse function α and the shift vector β^i. According to equation (2.39), the proper time τ as measured by a normal observer (whose four-velocity is n^a) increases by αdt when the coordinate time advances by dt. According to equation (2.41), the normal observer will also see the spatial coordinates x^i change by $-\beta^i dt$. A coordinate observer, whose spatial coordinate labels x^i remain constant, moves on a worldline whose tangent is t^a and which is shifted by $\beta^i dt$ with respect to the normal observer.

choose a time coordinate.[8] We still have the freedom to choose our spatial coordinates, however, and that choice affects the "upstairs", contravariant components of the normal vector. Specifically, note that the normalization condition (2.35) is automatically satisfied by

$$n^a = \frac{1}{\alpha}(1, -\beta^i) \qquad (2.40)$$

for *any* spatial vector β^i. Viewing n^a as a four-velocity helps us to interpret the role of β^i. For $\beta^i = 0$ the spatial components of the normal observer's four-velocity vanish, meaning that the spatial coordinates x^i of the normal observer do not change; we can think of the spatial coordinates as being attached to the normal observer. For non-zero β^i, however, the normal observer moves with respect to the spatial coordinates. In other words, the spatial coordinate labels arbitrarily attached to a normal observer change as the observer moves through spacetime, and these labels move away from the observer with three-velocity β^i. Using an argument similar to (2.38) above, except that we

---

[8] Geometrically, the quantity n_a actually refers to the spatial slice, or hypersurface, itself, and is called a "one-form", while the quantity n^a is the "normal vector" to the slice (see also Appendix A). We will not make use of this distinction in this volume but will simply treat the quantities A_a and A^a as covariant ("downstairs") or contravariant ("upstairs") versions of the same object, and similarly for other tensors.

now measure changes in the functions x^i rather than in t, we find that normal observers measure the changes in the spatial coordinates as

$$dx^i = \left(\frac{\partial x^i}{\partial x^a}\right)(n^a d\tau) = \delta^i{}_a(n^a d\tau) = n^i d\tau = -\beta^i dt \qquad (2.41)$$

during an advance $d\tau$ in the proper time (see Fig. 2.2 for an illustration). Conversely, a *coordinate observer*, i.e. an observer who is attached to the spatial coordinates, appears to be shifted by $\beta^i dt$ with respect to the normal observer. We therefore call β^i the *shift vector*.

It is sometimes useful to define the *time vector*

$$t^a = \alpha n^a + \beta^a = (1,0,0,0), \qquad (2.42)$$

where $\beta^t = 0$; this vector is tangent to a coordinate observer's worldline.

> **Exercise 2.4** Return to exercise 2.3 and find the contravariant components n^a of the normal vector in the unprimed frame. Compare your result with (2.40) to identify the shift vector β^i. *Hint:* Use $n^a = (\partial x^a/\partial x^{b'})n^{b'}$.

2.2.4 The 3+1 Split of Electrodynamics

Now that we have split our spacetime into space and time, we will decompose tensorial objects that live in this spacetime in a similar way. In Section 2.2.2 we simply picked out the time and space components when we decomposed the tensors and equations. Now we need the projections of tensors either onto the spatial slices or along the normal vector. For example, we now decompose the electromagnetic spacetime vector A^a according to

$$A^a = A^a_{\parallel} + A^a_{\perp}, \qquad (2.43)$$

where $A^a_{\parallel}$ is parallel to the normal vector n^a and $A^a_{\perp}$ is perpendicular to the normal vector, i.e. tangent to our spatial slice. The magnitude of $A^a_{\parallel}$ is given by the projection of A^a along n^a,

$$\phi \equiv -n_b A^b. \qquad (2.44)$$

We may interpret n^a as the four-velocity of the normal observer, and we see that this ϕ is proportional to the time component of A^a, i.e. the gauge variable ϕ (see (2.27)), as seen by a normal observer. We then write

$$A^a_{\parallel} = \phi n^a = -n^a n_b A^b, \qquad (2.45)$$

so that $A_{\parallel}^a$ indeed points in the normal direction n^a. Now we can solve (2.43) for $A_{\perp}^a$ to find

$$A_{\perp}^a = A^a - A_{\parallel}^a = g^a_{\ b}A^b + n^a n_b A^b = (g^a_{\ b} + n^a n_b)A^b, \qquad (2.46)$$

where the mixed-index metric $g^a_{\ b}$ always acts like a Kronecker delta, $g^a_{\ b} = g^{ac}g_{cb} = \delta^a_{\ b}$. The term in parentheses is aptly called the *projection operator*,

$$\gamma^a_{\ b} \equiv g^a_{\ b} + n^a n_b, \qquad (2.47)$$

since it projects a spacetime tensor A^a into a slice perpendicular to n^a, i.e. into a spatial slice. Acting on every index of a general spacetime tensor it will result in a new tensor that is tangent to the spatial slice, and hence spatial. Collecting results we now have

$$A^a = \phi n^a + A_{\perp}^a. \qquad (2.48)$$

Exercise 2.5 (a) Show that $A_{\parallel}^a$ is indeed normal by showing that its contraction with the projection operator $\gamma^a_{\ b}$ vanishes.

(b) Show that $A_{\perp}^a$ is indeed spatial by showing that its contraction with the normal vector n^a vanishes.

(c) Show that

$$\gamma^a_{\ b}\gamma^b_{\ c} = \gamma^a_{\ c}. \qquad (2.49)$$

This property guarantees that projecting an object that has already been projected will no longer change the object. Putting it differently, the projection operator acts like the identity operator for spatial objects.

Exercise 2.6 Find the components of the operator $\gamma^a_{\ b}$ that projects into the $t = const$ slices described in exercises 2.3 and 2.4.

Before proceeding we will consider the projections of a few more tensors, since it will allow us to make an early acquaintance with objects that we will encounter again in the context of general relativity. Admittedly, though, this involves a few somewhat tedious calculations. To help with the process, here is a roadmap to objects that we will meet along the way and that will be important later on:

- We define the *extrinsic curvature* K_{ab} in (2.53); it measures how much a spatial slice is "warped" in the enveloping spacetime.
- We define the *spatial covariant derivative* D_a in (2.65); as the name suggests, it is the spatial analogue of the four-dimensional covariant derivative.

- We meet the *Lie derivative* in (2.73); it measures those changes in a tensor field that are not the result of a coordinate transformation.
- Finally, we derive (2.88), which decomposes spacetime derivatives of A^a into spatial derivatives of $A^a_\perp$ plus terms to make up for the difference. We will see that the four-dimensional Riemann tensor is related to its three-dimensional counterpart in a very similar way.

All the above will play important roles in our treatment of general relativity. Readers eager to move forward could just take a quick glance at the objects listed above, skip the somewhat lengthy derivations, and then move on to Section 2.3. Readers interested in learning how to manipulate some of the mathematical quantities ("index gymnastics"), however, may want to work through the remainder of this section first.

The projection operator is like a new tool, and we can apply it to a number of different objects. We start with a decomposition of the charge density four-vector j^a,

$$j^a = g^a{}_b j^b = (\gamma^a{}_b - n^a n_b) j^b = n^a(-n_b j^b) + \gamma^a{}_b j^b. \tag{2.50}$$

Note that we inserted the mixed-index metric $g^a{}_b = \delta^a{}_b$ in the first step. It acts like a Kronecker delta but can be rewritten in terms of $\gamma^a{}_b - n^a n_b$. Identifying $\rho = -n_b j^b$ as the charge density observed by a normal observer n^a and $j^a_\perp = \gamma^a{}_b j^b$ as the charge-current density, again as observed by a normal observer, we find

$$j^a = \rho n^a + j^a_\perp. \tag{2.51}$$

For rank-2 tensors we project each index individually, so that in general we can induce completely normal, completely spatial, or mixed projections. As an important example we will work out the projections of the covariant derivative of the normal vector, $\nabla_a n_b$. Essentially, this tensor measures the rate at which the b-component of the normal vector changes in the a-direction; therefore it is a rank-2 tensor. Since changes in the normal vector are related to how much the spatial slices are "warped", as illustrated in Fig. 2.3, the tensor contains important geometrical information about our slices.

Using the same trick as before, but now using a Kronecker delta $\delta^c{}_a = g^c{}_a$ for each index, we obtain

$$\nabla_a n_b = g^c{}_a g^d{}_b \nabla_c n_d = (\gamma^c{}_a - n^c n_a)(\gamma^d{}_b - n^d n_b)\nabla_c n_d$$

$$= \gamma^c{}_a \gamma^d{}_b \nabla_c n_d - \gamma^d{}_b n_a n^c \nabla_c n_d - \gamma^c{}_a n_b n^d \nabla_c n_d + n_a n_b n^c n^d \nabla_c n_d. \tag{2.52}$$

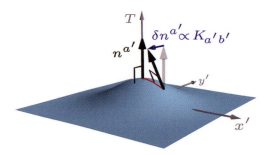

Figure 2.3 The extrinsic curvature K_{ab} measures how much a hypersurface is warped by computing differences between normal vectors at neighboring points. The covariant derivative of the normal vector field $n^{a'}$ along the red curve pointed in the direction $\mathbf{e}_{b'}$ measures departures from parallel transport. A non-zero departure, shown as the blue vector $\delta n^{a'}$, indicates that the surface is warped. The extrinsic curvature $K_{a'b'}$ is computed from the spatial projection of this covariant derivative.

The first term on the right-hand side of (2.52) is the complete spatial projection of $\nabla_a n_b$. It measures how much the normal vector changes as we move from point to point within one spatial slice. This quantity plays a crucial role in the context of general relativity also; we therefore define the *extrinsic curvature* as

$$K_{ab} \equiv -\gamma_a^c \gamma_b^d \nabla_c n_d, \qquad (2.53)$$

where the minus sign follows the convention that we will adopt.

Exercise 2.7 Insert the definition (2.34) into (2.53) to show that the extrinsic curvature is symmetric,

$$K_{ab} = K_{ba}, \qquad (2.54)$$

or equivalently $K_{[ab]} = 0$.

Exercise 2.8 Compute the extrinsic curvature for the $t = const$ slices in exercises 2.3, 2.4, and 2.6.

(a) Start by showing that the metric $g_{ab} = (\partial x^{c'}/\partial x^a)(\partial x^{d'}/\partial x^b)\,\eta_{c'd'}$ in the unprimed coordinate system is

$$g_{ab} = \begin{pmatrix} -1 & -h' & 0 & 0 \\ -h' & 1-(h')^2 & 0 & 0 \\ 0 & 0 & r^2 & 0 \\ 0 & 0 & 0 & r^2\sin^2\theta \end{pmatrix}. \qquad (2.55)$$

(b) Invert your result from part (a) to show that the inverse metric g^{ab} in the unprimed coordinate system is

$$g^{ab} = \begin{pmatrix} -1 + (h')^2 & -h' & 0 & 0 \\ -h' & 1 & 0 & 0 \\ 0 & 0 & r^{-2} & 0 \\ 0 & 0 & 0 & r^{-2}\sin^{-2}\theta \end{pmatrix}. \tag{2.56}$$

(c) Now expand the covariant derivative in (2.53),

$$\nabla_a n_b = \partial_a n_b - n_c \,^{(4)}\Gamma^c_{ab} = \partial_a n_b - n_t \,^{(4)}\Gamma^t_{ab}. \tag{2.57}$$

Evidently we need all the Christoffel symbols with a t index upstairs. Evaluate (1.31) to find all of them. *Hint:* The only non-zero ones are $^{(4)}\Gamma^t_{rr}$, $^{(4)}\Gamma^t_{\theta\theta}$, and $^{(4)}\Gamma^t_{\varphi\varphi}$; for example

$$^{(4)}\Gamma^t_{\theta\theta} = \frac{1}{2}g^{ta}(\partial_\theta g_{a\theta} + \partial_\theta g_{\theta a} - \partial_a g_{\theta\theta}) = \frac{1}{2}g^{tr}(\partial_\theta g_{r\theta} + \partial_\theta g_{\theta r} - \partial_r g_{\theta\theta}) = rh'. \tag{2.58}$$

(d) Now insert your results into (2.53) to find the extrinsic curvature K_{ij}. *Check:* $K_{\theta\theta} = -\alpha r h'$.

The term

$$a_a \equiv n^b \nabla_b n_a \tag{2.59}$$

measures the *acceleration* of a normal observer. For normal observers who follow geodesics, equation (1.28) implies $a_a = 0$, as expected. Finally, the normalization of the normal vector, $n_a n^a = -1$, implies that

$$n^a \nabla_b n_a = 0, \tag{2.60}$$

so that the last two terms on the right-hand side of (2.52) vanish. We can use the same argument to show that the acceleration a_a is purely spatial, $n^a a_a = 0$, so that $\gamma^b_a a_b = a_a$. The decomposition (2.52) of $\nabla_a n_b$ therefore reduces to

$$\nabla_a n_b = -K_{ab} - n_a a_b, \tag{2.61}$$

where the first term on the right-hand side is purely spatial, and the second term is a mixed spatial–normal projection.

We next consider projections of the Faraday tensor:

$$F^{ab} = g^a_{\ c} g^b_{\ d} F^{cd} = (\gamma^a_{\ c} - n^a n_c)(\gamma^b_{\ d} - n^b n_d) F^{cd}$$

$$= \gamma^a_{\ c}\gamma^b_{\ d}F^{cd} - \gamma^b_{\ d}n^a n_c F^{cd} - \gamma^a_{\ c}n^b n_d F^{cd} + n^a n^b n_c n_d F^{cd}. \tag{2.62}$$

Because F^{ab} is antisymmetric, the contraction $n_c n_d F^{cd}$ in the last term must vanish identically. In order to simplify the middle two terms we evaluate the Faraday tensor (2.18) in the normal observer's frame, i.e. with $u^a = n^a$, to find

$$E^a = n_b F^{ab} \tag{2.63}$$

(see exercise 2.2). The first term in (2.62), the completely spatial projection of the Faraday tensor, requires a little more work, but it also reduces to something very compact. Using the decomposition (2.48) we find

$$
\begin{aligned}
\gamma^a_c \gamma^b_d F^{cd} &= \gamma^a_c \gamma^b_d (\nabla^c A^d - \nabla^d A^c) \\
&= \gamma^a_c \gamma^b_d \left(\nabla^c (\phi n^d + A^d_\perp) - \nabla^d (\phi n^c + A^c_\perp) \right) \\
&= \gamma^a_c \gamma^b_d \left(n^d \nabla^c \phi - n^c \nabla^d \phi + \phi \nabla^c n^d - \phi \nabla^d n^c \right. \\
&\qquad\qquad \left. + \nabla^c A^d_\perp - \nabla^d A^c_\perp \right).
\end{aligned}
\tag{2.64}
$$

The first two terms vanish because the spatial projection of the normal vector vanishes, $\gamma^a_c n^c = 0$. The middle two terms can be rewritten in terms of the extrinsic curvature (2.53); we then recognize that these two terms cancel each other because the extrinsic curvature is symmetric (see 2.54). That leaves only the last two terms, which present us with a new type of object, namely the spatial projection of the covariant derivative of a spatial tensor. We define this as the *spatial covariant derivative* and denote it with the operator D_a, e.g.

$$D_a A^b_\perp \equiv \gamma^c_a \gamma^b_d \nabla_c A^d_\perp. \tag{2.65}$$

In Section 2.3 below we will discuss this derivative in more detail and will explain how it can be computed conveniently from spatial objects alone. For now it suffices to say that we define the spatial covariant derivative of a spatial tensor as the completely spatial projection of the tensor's spacetime covariant derivative. Using the definition (2.65) in (2.64) we now have

$$\gamma^a_c \gamma^b_d F^{cd} = D^a A^b_\perp - D^b A^a_\perp. \tag{2.66}$$

Collecting terms, we see that the decomposition (2.62) of the Faraday tensor reduces to

$$F^{ab} = D^a A^b_\perp - D^b A^a_\perp + n^a E^b - n^b E^a. \tag{2.67}$$

Exercise 2.9 Show that a normal observer measures the magnetic field B^a to be

$$B^a = \epsilon^{abc} D_b A_c^{\perp},\tag{2.68}$$

where $\epsilon^{abc} = n_d \epsilon^{dabc}$ is the spatial Levi–Civita tensor. *Hint:* Use part (b) of exercise 2.2.

Exercise 2.10 Show that the acceleration of a normal observer (2.59) is related to the lapse α according to

$$a_a = D_a \ln \alpha.\tag{2.69}$$

Hint: This requires several steps. Insert $n_a = -\alpha \nabla_a t$ into the definition of a_a, commute the second derivatives of t, use $n^b \nabla_a n_b = 0$, and finally recall the definition of the spatial covariant derivative.

In (2.67), we have used the electric field (2.63) to express a normal projection of the Faraday tensor. We now express this normal projection in terms of the vector potential A^a in order to obtain, in equation (2.82) below, an equation relating E_a to the time derivative of $A_a^{\perp}$. We start with

$$
\begin{aligned}
n^c F_{cd} &= n^c (\nabla_c A_d - \nabla_d A_c) = n^c \left(\nabla_c (\phi n_d + A_d^{\perp}) - \nabla_d (\phi n_c + A_c^{\perp}) \right) \\
&= n^c \left(\phi \nabla_c n_d + n_d \nabla_c \phi + \nabla_c A_d^{\perp} - \phi \nabla_d n_c - n_c \nabla_d \phi - \nabla_d A_c^{\perp} \right) \\
&= \phi a_d + n^c n_d \nabla_c \phi + \nabla_d \phi + n^c \nabla_c A_d^{\perp} + A_c^{\perp} \nabla_d n^c,
\end{aligned}
\tag{2.70}
$$

where we have used the acceleration (2.59) to rewrite the first term and have used (2.60) to eliminate the fourth term in the second line. In the last term we used the fact that $A_c^{\perp}$ is spatial, i.e. $n^c A_c^{\perp} = 0$, so that

$$-n^c \nabla_d A_c^{\perp} = A_c^{\perp} \nabla_d n^c.\tag{2.71}$$

Using (2.69) we can now combine the first three terms in (2.70):

$$
\begin{aligned}
\phi a_d + n^c n_d \nabla_c \phi + \nabla_d \phi &= \phi D_d \ln \alpha + (g_d^c + n^c n_d) \nabla_c \phi \\
&= \phi D_d \ln \alpha + \gamma_d^c \nabla_c \phi \\
&= \frac{1}{\alpha} (\phi D_d \alpha + \alpha D_d \phi) = \frac{1}{\alpha} D_d (\alpha \phi), \quad (2.72)
\end{aligned}
$$

where we have used $D_d \phi = \gamma_d^c \nabla_c \phi$. Finally, the trained eye may recognize the last two terms in (2.70) as the *Lie derivative* of $A_d^{\perp}$ along n^a,

$$\mathcal{L}_n A_d^{\perp} = n^c \nabla_c A_d^{\perp} + A_c^{\perp} \nabla_d n^c.\tag{2.73}$$

Whereas the covariant derivative of a vector A^a along another vector n^a vanishes if A^a is parallel-transported along n^a (see Appendix A.3), the Lie derivative along n^a vanishes if the changes in A^a result merely from an infinitesimal coordinate transformation generated by n^a.[9] Accordingly, the Lie derivative measures those changes in the tensor field that are not produced by a coordinate transformation generated by n^a.[10] Since Maxwell's equations govern physical changes in the dynamical fields, rather than coordinate changes, it is not too surprising to encounter the Lie derivative in this context. Quite in general, the Lie derivative plays an important role in formulating field equations in 3+1 decompositions, and it will reappear when we discuss general relativity in the next section.

Exercise 2.11 In general, the Lie derivative along a vector field w^a of a scalar ψ is

$$\mathcal{L}_w \psi = w^b \nabla_b \psi, \tag{2.74}$$

that of a contravariant vector field v^a is the *commutator* $[w^b, v^a]$,

$$\mathcal{L}_w v^a = [w^b, v^a] = w^b \nabla_b v^a - v^b \nabla_b w^a, \tag{2.75}$$

and that of a covariant vector field v_a is

$$\mathcal{L}_w v_a = w^b \nabla_b v_a + v_b \nabla_a w^b. \tag{2.76}$$

Use (2.74) and (2.75) to derive (2.76).

Exercise 2.12 Show that all terms involving Christoffel symbols in (2.73) cancel, so that we may also write the Lie derivative in terms of partial derivatives:

$$\mathcal{L}_n A_d^{\perp} = n^c \partial_c A_d^{\perp} + A_c^{\perp} \partial_d n^c. \tag{2.77}$$

This is a general property of Lie derivatives.

Among the properties of the Lie derivative is the following: taking the Lie derivative of any spatial tensor along αn^a results in a tensor that is again spatial. We therefore want to take Lie derivatives along αn^a rather than n^a in order to ensure that our spatial vectors E^a and $A_a^{\perp}$ remain spatial when evolved. Towards that end, we write

$$\alpha n^a = t^a - \beta^a, \tag{2.78}$$

---

[9] An infinitesimal coordinate transformation is generated by $x^{a'} = x^a + \delta\lambda\, n^a$, where λ parameterizes points on a curve to which the vector field $n^a = n^a(x^b)$ is tangent.

[10] See, e.g., Appendix A in Baumgarte and Shapiro (2010) for a detailed discussion of the Lie derivative.

where $\beta^a = (0, \beta^i)$, and where $t^a = (1, 0, 0, 0)$ is the time vector, which is tangent to a coordinate observer's worldline (see 2.42). We can then rewrite the Lie derivative (2.73) as

$$\alpha \mathcal{L}_n A_d^\perp = \mathcal{L}_{\alpha n} A_d^\perp = \mathcal{L}_t A_d^\perp - \mathcal{L}_\beta A_d^\perp. \tag{2.79}$$

Exercise 2.13 Verify equation (2.79).

From (2.77) we see that the Lie derivative along a coordinate vector such as t^a reduces to a partial derivative

$$\mathcal{L}_t A_d^\perp = \partial_t A_d^\perp. \tag{2.80}$$

Collecting terms we can now write the normal projection of the Faraday tensor as

$$n^c F_{cd} = \frac{1}{\alpha} \left(D_d(\alpha\phi) + \partial_t A_d^\perp - \mathcal{L}_\beta A_d^\perp \right). \tag{2.81}$$

According to (2.63) the left-hand side must be equal to $-E_d$, and we therefore find that

$$\partial_t A_a^\perp = -\alpha E_a - D_a(\alpha\phi) + \mathcal{L}_\beta A_a^\perp. \tag{2.82}$$

This equation is the generalization of (2.14) and (2.30) for arbitrary time slices and reduces to the latter for $\alpha = 1$ and $\beta^i = 0$. Crudely speaking, we see that we may interpret E^a as the time derivative of $A_a^\perp$.

As we have seen before, the evolution equation for the spatial components of the vector potential A^a, as well as the constraint equation, result from the divergence of the Faraday tensor, either (2.19) or, in terms of the vector potential, (2.26). Therefore our next task is to decompose the divergence of the Faraday tensor into spatial and normal parts. Using (2.67) we obtain

$$\nabla_a F^{ab} = \gamma^b_c \left(\mathcal{L}_n E^c - E^c K + \nabla_a(D^a A_\perp^c - D^c A_\perp^a) \right) - n^b D_a E^a, \tag{2.83}$$

where $\mathcal{L}_n E^c = n^a \nabla_a E^c - E^a \nabla_a n^c$ and where $K = g^{ab} K_{ab} = -\nabla_a n^a$ is the trace of the extrinsic curvature, also known as the *mean curvature*.

Exercise 2.14 Derive equation (2.83). *Hint:* As before, you can start with $\nabla_a F^{ab} = g^b_c \nabla_a F^{ac} = (\gamma^b_c - n^b n_c) \nabla_a F^{ac}$, then insert (2.67) in to the first term, make the identifications $E^a = n_c F^{ac}$ as well as $\nabla_a n_c = -K_{ac} - n_a a_c$ (see 2.61) in the second term, and show that $D_a E^a = \nabla_a E^a - E^a a_a$.

We now want to work on the third term on the right-hand side of (2.83). Just to make the notation a little easier, we will temporarily introduce $S^{ab} \equiv (D^a A^c_\perp - D^c A^a_\perp)$, so that this term becomes $\gamma^b_c \nabla_a S^{ab}$. We can then convert the covariant spacetime derivative ∇_a into a covariant spatial derivative D_a by inserting two copies of the Kronecker delta $\delta^a_d = g^a_d = \gamma^a_d - n^a n_d$, which yields

$$\gamma^b_c \nabla_a S^{ac} = \gamma^b_c (\gamma^a_d - n^a n_d)(\gamma^e_a - n^e n_a) \nabla_e S^{dc}$$

$$= \gamma^b_c \gamma^a_d \gamma^e_a \nabla_e S^{dc} + \gamma^b_c n^a n_d n^e n_a \nabla_e S^{dc}. \tag{2.84}$$

We were able to cancel several terms in the last equality because $\gamma^a_d n_a = 0$. Since S^{dc} is purely spatial, we recognize the first term in the last equality as the covariant spatial derivative (see 2.65). In the second term we write $n_d \nabla_e S^{dc} = -S^{dc} \nabla_e n_d$, which is again a consequence of S^{dc} being purely spatial, so that $n_d S^{dc} = 0$ (compare 2.71). Further, using $n^a n_a = -1$ we obtain

$$\gamma^b_c \nabla_a S^{ab} = D_a S^{ab} + \gamma^b_c S^{dc} n^e \nabla_e n_d. \tag{2.85}$$

Finally, we note that $\gamma^b_c S^{dc} = (\delta^b_c + n^b n_c) S^{dc} = S^{db}$, use the definition of the normal observer's acceleration (2.59), and reinsert our definition of S^{ab} to find

$$\gamma^b_c \nabla_a (D^a A^c_\perp - D^c A^a_\perp) = D_a (D^a A^b_\perp - D^b A^a_\perp) + (D^a A^b_\perp - D^b A^a_\perp) a_a. \tag{2.86}$$

Using (2.69) we can also rewrite (2.86) in the more compact form

$$\gamma^b_c \nabla_a (D^a A^c_\perp - D^c A^a_\perp) = \alpha^{-1} D_a (\alpha D^a A^b_\perp - \alpha D^b A^a_\perp). \tag{2.87}$$

Collecting terms we may therefore rewrite (2.83) as

$$\nabla_a F^{ab} = \nabla_a (\nabla^a A^b - \nabla^b A^a) = \alpha^{-1} D_a (\alpha D^a A^b_\perp - \alpha D^b A^a_\perp)$$

$$+ \gamma^b_a \mathcal{L}_n E^a - E^b K - n^b D_a E^a, \tag{2.88}$$

where the first three terms in (2.88) are purely spatial, and the last term is normal. We can also expand the Lie derivative term as

$$\alpha \gamma^b_c \mathcal{L}_n E^c = \gamma^b_c \mathcal{L}_{\alpha n} E^c = \partial_t E^b - \mathcal{L}_\beta E^b. \tag{2.89}$$

Exercise 2.15 Derive equation (2.89).

It is well worth inspecting equation (2.88) a little more carefully. The second expression has four-dimensional derivatives of our four-dimensional fundamental quantity, A^a. The first term in the third expression has a similar appearance, except that it involves spatial derivatives of the spatial projection of this fundamental quantity only. It is evident that this term alone cannot contain all the terms on the left-hand side. For example, the second expression contains up to second time derivatives of the fundamental variable, which cannot be accommodated by the first term in the third expression. These missing terms are accounted for by the remaining terms in the third expression – the second time derivatives, for example, appear in the term $\partial_t E^b$ by virtue of (2.82).

We also note that we have not invoked Maxwell's equations in our derivation of (2.88) – instead, equation (2.88) results from the decomposition of the vector field A^a and its derivatives in terms of spatial objects. Stated differently, equation (2.88) is a consequence of the 3+1 split of spacetime and is independent of physical laws. We can now obtain Maxwell's equations in a 3+1 decomposition, though, by inserting (2.88) into the field equation (2.26) and considering the spatial and normal parts separately. Specifically, we see that the spatial projection of that equation now yields the evolution equation:

$$\partial_t E^b = D_a(\alpha D^b A_\perp^a - \alpha D^a A_\perp^b) + \mathcal{L}_\beta E^b + \alpha E^b K - 4\pi\alpha j_\perp^b, \quad (2.90)$$

which generalizes our earlier result (2.15). The normal projection of (2.88) results in the constraint equation,

$$D_a E^a = 4\pi\rho, \quad (2.91)$$

which, in fact, takes a form identical to (2.16). In (2.91) we have also reintroduced our color-coding from the previous sections. Doing the same for (2.82) and (2.90) we obtain

$$\partial_t A_a^\perp = -\alpha E_a - D_a(\alpha\phi) + \mathcal{L}_\beta A_a^\perp, \quad (2.92)$$

$$\partial_t E^a = D_b(\alpha D^a A_\perp^b - \alpha D^b A_\perp^a) + \alpha E^a K - 4\pi\alpha j_\perp^a + \mathcal{L}_\beta E^a. \quad (2.93)$$

This pair of equations now generalizes (2.14) and (2.15) for general time slices, i.e. for general 3+1 splittings of the spacetime.[11] Once we have computed $A_a^\perp$, we can also compute the magnetic field, as observed by a normal observer, from (2.68).

Exercise 2.16 Retrace the steps of this section to consider the scalar field of Section 2.1 for arbitrary time slices.
(a) Define $\kappa \equiv -n^a \nabla_a \psi$ and show that this may be written as

$$\partial_t \psi = -\alpha\kappa + \mathcal{L}_\beta \psi, \tag{2.94}$$

where we have used $\mathcal{L}_\beta \psi = \beta^i \partial_i \psi$ for a scalar field.
(b) Show that decomposition of the spacetime gradient $\nabla_a \psi$ yields

$$\nabla_a \psi = D_a \psi + n_a \kappa. \tag{2.95}$$

(c) Now use steps similar to those in exercise 2.14 to construct the decomposition of $\nabla_a \nabla^a \psi = g_a^{\ b} \nabla_b \nabla^a \psi$. Replace $\nabla^a \psi$ with (2.95), insert (2.47) for $g_a^{\ b}$, and then use the definition of the extrinsic curvature and the spatial covariant derivative (together with 2.49), as well as properties of the derivatives of the normal vector n_a, to show that

$$\partial_t \kappa = -D_a(\alpha D^a \psi) + \alpha\kappa K + 4\pi\alpha\rho + \mathcal{L}_\beta \kappa. \tag{2.96}$$

Note the remarkable similarity between the pair of equations (2.93) and (2.92) for electrodynamics, and (2.94) and (2.96) for the scalar field.

In principle, we could now solve the initial-value or Cauchy problem for electrodynamics as follows. We first choose initial data E^a and $A_a^\perp$, where the former have to satisfy the constraint equation (2.91). We also need to make a choice for ϕ, which determines the gauge for the electromagnetic fields as well as for the lapse α and the shift β^i, which in turn determine the spacetime coordinates. With these choices we can then integrate equations (2.92) and (2.93) forward in time, thereby obtaining solutions for E^a and $A_a^\perp$ for all times. Combining the latter with $A_{\parallel}^a = \phi n^a$ yields the four-vector A^a, from which we can construct the Faraday tensor as in (2.25). If desired, we can also compute the corresponding magnetic field by identifying the corresponding components of the Faraday tensor, as in (2.17) and exercise 2.2, or by using the relation $B^i = \epsilon^{ijk} D_j A_k$. The observant reader will notice, however, that some of the above steps also require knowledge of the spatial metric, which we have to construct on each

[11] An alternative 3+1 split is given by Thorne and MacDonald (1982). It provides evolution equations for E^i and B^i and requires the familiar constraint equations for each of these variables.

slice along with the electromagnetic fields. Doing just that – evolving the metric in a 3+1 decomposition – will be the subject of the next section. In fact, the equations that we have derived in this section for scalar fields and electrodynamics apply even in the presence of gravitational fields, to which we are about to return.

To summarize: at the beginning of Section 2.2.3 we introduced a function t whose level surfaces we identified with our spatial hyper-surfaces. Given this function, we then introduced the normal vector to those slices, and expressed its components in terms of the lapse α and the shift β^i. Ultimately, we brought Maxwell's equations into the form (2.92) and (2.93). After that, we can freely choose the lapse and the shift and thereby specify our time and space coordinates. We often say that the lapse and shift encode our coordinate freedom. In other words, imagine laying down arbitrary spatial coordinates on a spacelike hypersurface labeled by the parameter t. When normal observers move to a new hypersurface labeled by $t + dt$, they will find that their watches will have advanced by a proper time αdt and their spatial coordinate labels have shifted an amount $-\beta^i dt$, as shown in Fig. 2.2.

Determining proper distances between spatial coordinates on the new hypersurface requires more than the lapse and the shift. In the next section we will encounter the spatial metric, the projection of the spacetime metric onto spatial hypersurfaces, which is tailored for making these measurements. We will also see how this spatial metric evolves from time slice t to time slice $t + dt$, together with all other fields.

2.3 The 3+1 Split of General Relativity

Finally we are ready to discuss a 3+1 decomposition of Einstein's gravitational field equations. We are indeed ready, since we introduced almost all the necessary objects and concepts in Section 2.2. There, we discussed the 3+1 split of spacetime and proceeded to decompose Maxwell's equation (2.26) for the vector potential A^a; now we need to decompose Einstein's equation (1.37) for the spacetime metric g_{ab}. Similarly to how Maxwell's equation (2.26) appears, at first sight, to provide four equations for the four unknowns in A^a, Einstein's equation (1.37) appears to provide ten equations for the ten independent components in g_{ab}. Recall, though, that the four components in (2.26) are not independent, since the divergence of the equations vanishes by virtue of the "electromagnetic Bianchi identity" (2.22). We concluded that one equation in (2.26) must be redundant, leading to the appearance of both a constraint equation and a gauge freedom. Not surprisingly, we encounter a very similar structure in general relativity.

In Section 1.2.4 we mentioned that the Einstein tensor (1.36) is the only viable combination of the Ricci tensor and scalar that will automatically conserve the energy and momentum of the source of gravitation. To see this, we can take the covariant derivative of (1.36), use the fact that the covariant derivative of the metric always vanishes, $\nabla_a g_{bc} = 0$ (see the discussion in Appendix A.3), and employ the Bianchi identity (D.13) to find

$$\nabla_a G^{ab} = 0. \tag{2.97}$$

As this is the contracted version of the Bianchi identity (D.13), it is often referred to as the contracted Bianchi identity. By virtue of Einstein's equations (1.37) it implies that $\nabla_a T^{ab} = 0$, which encodes the local conservation of energy and momentum and yields the equations of motion for the matter fields. We demonstrate how this works for scalar and electromagnetic fields in Appendix C.

The Bianchi identity (2.97) play the same role in general relativity as (2.22) does in electrodynamics. Both equations state that the divergence of the left-hand side of the field equations vanishes; therefore they both imply conservation laws for the matter terms on the right-hand side. They also demonstrate that the individual equations in the field equations are not independent. Given that (2.97) is a vector equation (it has only one free index, namely b) with four components, it tells us that four of the ten components of Einstein's equations cannot be independent. We therefore expect that the constraint equations in general relativity will have a total of four components, and, by the same token, we expect a four-fold gauge freedom. We will identify this gauge freedom with the four-fold coordinate freedom in general relativity.

We can now proceed as in Section 2.2 – we will first decompose the variable and then the equation. Our variable is now the spacetime metric g_{ab}, which we decompose as

$$g_{ab} = g^c{}_a g^d{}_b g_{cd} = (\gamma^c_a - n^c n_a)(\gamma^d_b - n^d n_b)g_{cd} = \gamma^c_a \gamma^d_b g_{cd} - n_a n_b. \tag{2.98}$$

Notice that the first term on the right-hand side,

$$\gamma^c_a \gamma^d_b g_{cd} = g_{ab} + n_a n_b, \tag{2.99}$$

is the projection operator that we defined in (2.47), but here with both indices downstairs. We now refer to this object as the *spatial metric*

$$\boxed{\gamma_{ab} \equiv g_{ab} + n_a n_b.}$$
(2.100)

Drawing on the analogy with electromagnetism we note that the spatial metric γ_{ab} is the gravitational analogue of the spatial vector potential $A^a_\perp$.

Exercise 2.17 We can "count" the number of dimensions d of a space by taking the trace of the metric, i.e. $d = g^a_a$. Take the trace of the induced metric $\gamma_{ab} = g_{ab} + n_a n_b$ to show that the dimension of the spatial slice is one less than that of the spacetime.

The spatial metric plays the same role for spatial tensors as the spacetime metric does for spacetime tensors. In particular, we raise and lower the indices of spatial tensors with the spatial metric rather than the spacetime metric. For the shift vector, for example, we have

$$\beta_a = g_{ab}\beta^b = (\gamma_{ab} - n_a n_b)\beta^b = \gamma_{ab}\beta^b,$$
(2.101)

since $n_b\beta^b = 0$. As $n_a = (-\alpha, 0, 0, 0)$, this last equality reduces to $n_t\beta^t = 0$, showing that $\beta^t = 0$. In fact, the contravariant ("upstairs") time component of spatial tensors must always vanish. Therefore, we may restrict our indices to spatial indices when we are dealing with spatial tensors. As mentioned earlier, we will use the convention that $a, b, c, \ldots$ are spacetime indices and $i, j, k, \ldots$ are spatial indices only. As a specific example, the time components of the inverse spatial metric must vanish; we can therefore write

$$\boxed{g^{ab} = \gamma^{ab} - n^a n^b = \begin{pmatrix} -\alpha^{-2} & \alpha^{-2}\beta^i \\ \alpha^{-2}\beta^j & \gamma^{ij} - \alpha^{-2}\beta^i\beta^j \end{pmatrix}.}$$
(2.102)

We can invert this metric to find

$$\boxed{g_{ab} = \begin{pmatrix} -\alpha^2 + \beta_l\beta^l & \beta_i \\ \beta_j & \gamma_{ij} \end{pmatrix},}$$
(2.103)

where $\beta_l = \gamma_{la}\beta^a = \gamma_{lk}\beta^k$, or we can confirm this result by verifying that $g^{ac}g_{cb} = \delta^a_b$. The four-dimensional line element now becomes

$$ds^2 = g_{ab}dx^a dx^b = -\alpha^2 dt^2 + \gamma_{ij}(dx^i + \beta^i dt)(dx^j + \beta^j dt).$$
(2.104)

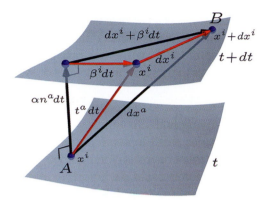

Figure 2.4 An extension of Fig. 2.2, illustrating the interpretation of the line element (2.104) as the "Pythagorean theorem" in (3+1)-dimensional spacetime. The vectors $\alpha n^a dt$ and $t^a dt$ connect points on two neighboring time slices. The vector $\beta^i dt$ resides in the slice and measures their difference. The infinitesimal total displacement vector dx^a connects two nearby, but otherwise arbitrary, points on neighboring slices (e.g. the point A at x^i on slice t and the point B at $x^i + dx^i$ on slice $t + dt$). The total displacement vector $dx^a = t^a dt + dx^i$, where dx^i is the spatial vector drawn in the figure, may be decomposed alternatively into two vectors that form the legs of the right-triangle drawn by the three black vectors, $dx^a = \alpha n^a dt + (dx^i + \beta^i dt)$. Using this decomposition to evaluate the invariant interval $ds^2 = dx_a dx^a = g_{ab} dx^a dx^b$ results in the Pythagorean theorem (2.104).

As the illustration in Fig. 2.4 suggests, we can think of this expression for ds^2 as a "spacetime Pythagorean theorem" for the invariant distance ds^2 between two points (t, x^i) and $(t + dt, x^i + dx^i)$: the first term on the right-hand side measures the proper time from t to $t + dt$ along the normal vector from one slice to the next (which we have already identified as αdt in 2.39), while the remaining terms measure the proper length from x^i to $x^i + dx^i$ within the spatial slice.

Notice that, whenever we are dealing with spatial tensors, we may restrict the indices to spatial indices only. For the equations of electrodynamics, for example, the constraint equation (2.91) becomes

$$D_i E^i = 4\pi \rho, \tag{2.105}$$

while the two evolution equations (2.92) and (2.93) reduce to

$$\partial_t A_i^\perp = -\alpha E_i - D_i(\alpha \phi) + \mathcal{L}_\beta A_i^\perp, \tag{2.106}$$

$$\partial_t E^i = D_j(\alpha D^i A_\perp^j - \alpha D^j A_\perp^i) + \alpha E^i K - 4\pi \alpha j_\perp^i + \mathcal{L}_\beta E^i. \tag{2.107}$$

Exercise 2.18 Show that the determinant g of the spacetime metric is related to the determinant of the spatial metric, γ, by

$$(-g)^{1/2} = \alpha(\gamma)^{1/2}. \tag{2.108}$$

Hint: First write $-\alpha^2 = g^{00}$ and then recall that for any square matrix A_{ij} the following is true: $(A^{-1})_{ij} = $ cofactor of $A_{ji}/\det A$.

It is a good exercise to compare the above expressions for the spacetime metric with known analytical solutions of Einstein's equations in order to identify the lapse function α, the shift vector β^i, and the spatial metric γ_{ij}. This is a little easier when the shift vector vanishes...

Exercise 2.19 If all went well in exercise 1.5, you should have found that the Schwarzschild metric in isotropic coordinates (on slices of constant Schwarzschild time) is given by

$$ds^2 = -\left(\frac{1 - M/(2r)}{1 + M/(2r)}\right)^2 dt^2 + \left(1 + \frac{M}{2r}\right)^4 (dr^2 + r^2 d\Omega^2). \tag{2.109}$$

Compare this line element with the "3+1" line element (2.104) to identify the lapse function α, the shift vector β^i, and the spatial metric γ_{ij} in spherical polar coordinates.

...and a little harder when it doesn't.

Exercise 2.20 Revisit the Schwarzschild solution in the trumpet coordinate system (1.48) of exercise 1.6.

(a) Compare the line elements (2.104) and (1.48) to identify $\beta_{\bar{r}} = \gamma_{\bar{r}\bar{r}}\beta^{\bar{r}}$ and the spatial metric $\gamma_{\bar{i}\bar{j}}$ on slices of constant time $\bar{t}$.

(b) Invert the spatial metric and compute $\beta^{\bar{r}}$.
(c) Compute the lapse function α.
Check: For $\bar{r} = M$ you should find $\alpha = 1/2$.

Exercise 2.21 Identify the lapse, the shift, and the spatial metric for the spacetime metric that you found in exercise 2.8, and verify that they agree with your results from exercises 2.3 and 2.4.

Next we will ask the reader to show that we can express the extrinsic curvature (2.53) as the Lie derivative of γ_{ab} along the normal vector n^a.

Exercise 2.22 Evaluate the Lie derivative of $\gamma_{ab} = g_{ab} + n_a n_b$ along n^a,

$$\mathcal{L}_n \gamma_{ab} = n^c \nabla_c \gamma_{ab} + \gamma_{cb} \nabla_a n^c + \gamma_{ac} \nabla_b n^c, \tag{2.110}$$

to show that

$$K_{ab} = -\frac{1}{2}\mathcal{L}_n \gamma_{ab}. \tag{2.111}$$

Hint: Recall (2.59) and (2.61), as well as $\nabla_a g_{bc} = 0$.

Using (2.78) and the discussion that follows it, we can rewrite
(2.111) as

$$\partial_t \gamma_{ij} = -2\alpha K_{ij} + \mathcal{L}_\beta \gamma_{ij} = -2\alpha K_{ij} + D_i \beta_j + D_j \beta_i. \qquad (2.112)$$

Evidently, we can think of the extrinsic curvature as measuring the time
derivative of the spatial metric.

In a full disclosure, two more properties of derivatives have gone
into the derivation of the last equation. First, we have again used
the fact that the covariant derivative is compatible with its associated
metric, i.e. $\nabla_a g_{bc} = 0$ and similarly $D_a \gamma_{bc} = 0$ (see the discussion
in Appendix A.3). Second, it turns out that the Lie derivative can
be computed using *any* covariant derivative.[12] In fact, evaluating the
covariant derivatives in (2.110) shows that all the Christoffel symbols
drop out identically, as in exercise 2.12. We can therefore replace the
covariant derivatives in (2.110) with partial derivatives, or, as we did in
deriving (2.112), with the spatial covariant derivatives. In that case the
term $D_c \gamma_{ab}$ vanishes, and we are left with spatial covariant derivatives
of the shift vector only.

In equation (2.112) we have also reintroduced the color-coding
of the previous sections, which helps one to see the similarity with
equations (2.4) and (2.14) (or 2.82). We have previously identified γ_{ij}
as the relativistic analogue of ψ for a scalar field or of $A_\perp^a$ for an
electromagnetic field; comparing (2.112) with (2.4) or (2.14) hints that
the intrinsic curvature K_{ab} plays the same role as κ for the scalar field
and as the electric field E^a in electromagnetism.

Here is a point of potential confusion. In electromagnetism in a fixed
spacetime, we assume that a metric is given.[13] In a four-dimensional
description, the vector potential A^a then evolves according to (2.26)
in this given spacetime. For example, if the spacetime is flat, we may
take the spacetime metric g_{ab} to be given by the Minkowski metric
η_{ab}. We can then introduce spatial slices, which result in our definition
(2.53) of the extrinsic curvature K_{ab}. Evidently, we can also define
the spatial metric γ_{ab} as in (2.100). In electromagnetism in a fixed

---

[12] Here and throughout we assume a coordinate basis and that the covariant derivative is
torsion-free.

[13] In general, the electromagnetic field affects the metric and its evolution through the
electromagnetic stress–energy tensor T_{EM}^{ab}. However, in many cases the gravitational effect of
the electromagnetic fields may be negligible. Unless there are other dynamical sources of
gravitational fields, we may then approximate the spacetime metric as given and fixed. For
example, in the absence of any gravity it would be the flat Minkowski metric ($g_{ab} = \eta_{ab}$) or
for very weak electromagnetic fields in the presence of a non-rotating black hole it would be
the Schwarzschild metric.

spacetime, however, neither γ_{ab} nor K_{ab} are dynamical variables; they are determined by the fixed spacetime metric and the slicing that has been adopted. Instead, the vector potential A^a is considered the dynamical variable, or, in a 3+1 decomposition, the spatial vectors $A^i_\perp$ and E^i are the dynamical variables.

In general relativity, the spacetime metric g_{ab} itself becomes the dynamical variable – therefore it simultaneously plays the role of the metric and of the vector potential A^a in electrodynamics. By the same token, the spatial metric γ_{ij} and extrinsic curvature K_{ij} become dynamical variables in a 3+1 decomposition of general relativity. When comparing with electromagnetism in a fixed spacetime, γ_{ij} therefore absorbs the dual roles of the spatial metric and $A^i_\perp$, while K_{ij} takes over the dual roles of extrinsic curvature and E^i.[14]

Having defined a spatial metric γ_{ij} that measures proper distances on spatial slices, we can retrace the outline of Section 1.2 and develop for the spatial metric γ_{ij} the same geometric objects that we have previously constructed for the spacetime metric g_{ab}.

First on the to-do list is the covariant derivative. We compute the covariant derivative associated with the spatial metric in exactly the same way as for the spacetime metric, except that we replace all terms with the corresponding spatial terms. For example, we have

$$D_i V^j = \partial_i V^j + V^k \Gamma^j_{ki}, \qquad (2.113)$$

where V^j is a spatial vector, and where the Christoffel symbols Γ^j_{kj}, the three-dimensional version of their spacetime cousins (1.31), are computed from derivatives of the spatial metric,

$$\Gamma^i_{jk} = \frac{1}{2}\gamma^{il}\left(\partial_k \gamma_{jl} + \partial_j \gamma_{lk} - \partial_l \gamma_{jk}\right). \qquad (2.114)$$

The observant reader will notice that we defined the spatial covariant derivative before, in (2.65), as the total spatial projection of the four-dimensional covariant derivative of a spatial tensor. As exercise 2.23

---

[14] Above, we have maintained separate coordinate and gauge freedoms in electromagnetism: coordinate freedom (i.e. the choice of slicing, together with the choice of how the spatial coordinates evolve from slice to slice) is expressed by α and β^i, while the gauge freedom is expressed by ϕ. By contrast, in general relativity we have chosen a basis that reflects the 3+1 foliation. Effectively, this attaches the gauge freedom to the slicing, so that we no longer have separate coordinate and gauge freedom (see also footnote 6 above, as well as Section 2.7 in Baumgarte and Shapiro (2010) for a detailed discussion). This is a very common approach, but for some applications it may be advantageous to adopt a *dual-frame* formalism that maintains the coordinate and gauge freedom separately; see, e.g., Scheel et al. (2006); Hilditch (2015).

demonstrates, the two ways of writing the spatial covariant derivative are completely consistent.

Exercise 2.23 Show that, for a spatial vector V^b,

$$\gamma_i^a \gamma_b^j \nabla_a V^b = \partial_i V^j + V^k \Gamma_{ki}^j. \tag{2.115}$$

Hint: Expand the left-hand side in terms of spacetime Christoffel symbols $^{(4)}\Gamma_{bc}^a$, then use $g_{ab} = \gamma_{ab} - n_a n_b$ in the latter, and show that all terms involving the normal vector drop out. Note that this argument relies on V^a being spatial, so that $n_a V^a = 0$.

Next on the to-do list is the Riemann tensor and then its trace, the Ricci tensor. We compute the spatial versions of these tensors in exactly the same way as before, except that we replace the spacetime metric with the spatial metric. For example, instead of (1.34) we now have $R_{ij} = R_{ikj}^k$, or

$$R_{ij} = -\frac{1}{2}\gamma^{kl}\left(\partial_k\partial_l\gamma_{ij} + \partial_i\partial_j\gamma_{kl} - \partial_i\partial_l\gamma_{kj} - \partial_k\partial_j\gamma_{il}\right)$$
$$+ \gamma^{kl}\left(\Gamma_{il}^m\Gamma_{mkj} - \Gamma_{ij}^m\Gamma_{mkl}\right), \tag{2.116}$$

where $\Gamma_{ijk} = \gamma_{il}\Gamma_{jk}^l$. Here is a key point: unlike the spatial covariant derivative, which we can compute from the spatial projection of its spacetime counterpart, the spatial Riemann tensor is *not* the spatial projection of the spacetime Riemann tensor, and neither is the spatial Ricci tensor R_{ij} the spatial projection of the spacetime Ricci tensor $^{(4)}R_{ab}$. Note that the spatial projection of the spacetime covariant derivative becomes the spatial covariant derivative *only* for spatial tensors – see exercise 2.23. In the spacetime Ricci tensor, however, we are taking derivatives of the spacetime metric, which certainly is not spatial. To illustrate this point, return to exercises 2.3, 2.4, 2.6, 2.8, and 2.21.

Exercise 2.24 (a) Find the spatial Christoffel symbols Γ_{jk}^i on the $t = const$ slices of exercise 2.3 (assuming you found $\gamma_{ij} = \mathrm{diag}(1-(h')^2, r^2, r^2\sin^2\theta)$ in exercise 2.21). *Hint:* There are seven non-zero Christoffel symbols, not counting those related by symmetry; one of them is $\Gamma_{rr}^r = -\alpha^2 h' h''$, where $h'' = d^2h/dr^2$. Note also that, for $h' = 0$, the Christoffel symbols should reduce to those for a flat metric in spherical coordinates; see equation (A.47).
(b) Find the spatial Ricci tensor R_{ij} on these slices. *Hint:* It is probably easiest to use expression (D.16) for the Ricci tensor here; one component is $R_{rr} = -2\alpha^2 h' h''/r$.

Let us appreciate the result of exercise 2.24. We started in exercise 2.3 with a flat spacetime since, in the primed coordinate system, we had $g_{a'b'} = \eta_{a'b'}$. If the spacetime is flat, its Ricci tensor must vanish. But the Ricci tensor is indeed a tensor, as the name suggests, so if it vanishes in the primed coordinate system then it must also vanish in the unprimed coordinate system. We therefore have $^{(4)}R_{ab} = 0$. Already, in exercise 2.8, where the spacetime was assumed flat, we found that the extrinsic curvature on the $t = const$ slices of exercise 2.3 is non-zero as long as h' is non-zero; exercise 2.24 now demonstrates that its intrinsic curvature, measured by R_{ij}, is also non-zero as long as h' does not vanish. Evidently, R_{ij} is not the spatial projection of $^{(4)}R_{ab}$.

Invoking the analogy with electromagnetism again, we should not be surprised by this observation. The electromagnetic analogue of the spacetime Ricci tensor is the term on the left-hand side of (2.88), while the analogue of the spatial Ricci tensor is the first term on the right-hand side. The former contains spacetime derivatives of the four-dimensional fundamental quantity, while the latter contains only spatial derivatives of the spatial projection of the fundamental quantity.[15] However, from (2.88) we also see that the spatial projection of the former is the latter plus extra terms. A projection of the spacetime Ricci tensor similarly includes both the spatial Ricci tensor and extra terms, and it is these extra terms that explain the difference between the spatial projections of $^{(4)}R_{ab}$ and R_{ij} (see equation D.27).

In fact, in order to obtain our desired 3+1 decomposition of Einstein's equations, we require not only the spatial projection of the Riemann tensor but a complete decomposition, as in (2.88). We could derive this decomposition by carrying out a program very similar to that in Section 2.2.3; alternatively, we could start with the definition (D.10) of the spacetime Riemann tensor and then express the four-dimensional covariant derivatives as spatial covariant derivatives plus extra terms. The results of this derivation are the equations of Gauss, Codazzi, Mainardi, and Ricci. We will not derive these equations here[16] but list them, for completeness, in Appendix D.4. Invoking the analogy with (2.88), however, we can anticipate what kind of terms we should expect even without a formal derivation.

---

[15] Note that, in the electromagnetic case, the spatial version includes some lapse function factors. These arise because the derivatives are covariant derivatives; see equation (2.86). The Ricci tensor involves partial derivatives only, and therefore these extra factors are absent.

[16] See, e.g., Section 2.5 in Baumgarte and Shapiro (2010) for a detailed treatment.

The first term on the right-hand side of (2.88) is the spatial counter-part of the left-hand side; in general relativity, this becomes the spatial Ricci tensor (2.116). The next term is the Lie derivative of E^a along n^a – by analogy, we now expect its counterpart to be the Lie derivative of the extrinsic curvature K_{ab} along n^a. As in electrodynamics (see equation 2.89) we can express this Lie derivative in terms of the time derivative of K_{ab} and the Lie derivative along β^i. The next term in (2.88) is the product $E^a K$. In general relativity, the extrinsic curvature takes over the role of both E^a and K, therefore we should not be surprised to find terms quadratic in K_{ab}. Finally, we expect to encounter the spatial divergence of K_{ab}, in analogy to the last term in (2.88).

Just as for (2.88), the equations of Gauss, Codazzi, Mainardi, and Ricci are based on differential geometry only, i.e. on the 3+1 decomposition of spacetime, and do not yet invoke Einstein's equations. In order to complete our 3+1 decomposition of Einstein's equations, we now equate the projections of the Einstein tensor G_{ab} with projections of the stress–energy tensor T_{ab} on the right-hand side of (1.37). Projecting both indices along the normal results in the *Hamiltonian constraint*

$$R + K^2 - K_{ij}K^{ij} = 16\pi\rho, \tag{2.117}$$

where

$$\rho \equiv n^a n^b T_{ab} \tag{2.118}$$

is the total mass–energy density as measured by a normal observer and $K = \gamma^{ij}K_{ij}$ is the trace of the extrinsic curvature, also known as the *mean curvature*. A mixed spatial–normal projection yields a second constraint equation, the *momentum constraint*

$$D_j(K^{ij} - \gamma^{ij}K) = 8\pi j^i, \tag{2.119}$$

where

$$j^i \equiv -\gamma^{ia}n^b T_{ab} \tag{2.120}$$

is the mass–energy flux, or momentum density, as seen by a normal observer. We have reintroduced our color coding in order to emphasize the similarity of equation (2.119) to the electrodynamic constraint equation (2.91) – a divergence of the "red" variable is equal to a matter term. Note that there are four independent components in the constraint equations (2.117) and (2.119), as we anticipated at the beginning of this section. We can think of these constraint equations as conditions on the internal and external curvatures of the spatial slices that guarantee that they will all "fit" into the curvature of the spacetime prescribed by its mass–energy content. To illustrate this, we return one more time to exercise 2.3 and its sequels.

> **Exercise 2.25** Evaluate R from your results in exercise 2.24, and K^2 and $K_{ij}K^{ij}$ from those of exercise 2.8, to show that the Ricci scalar and extrinsic curvature of the $t = const$ slices of exercise 2.3 satisfy the Hamiltonian constraint (2.117) with $\rho = 0$.

Finally, a complete spatial projection results in the evolution equation for the extrinsic curvature.[17] Combining this result with (2.112) we obtain the pair

$$\partial_t \gamma_{ij} = -2\alpha K_{ij} + \mathcal{L}_\beta \gamma_{ij}, \tag{2.121}$$

$$\partial_t K_{ij} = \alpha(R_{ij} - 2K_{ik}K^k_{\ j} + K K_{ij}) - D_i D_j \alpha - 4\pi\alpha M_{ij} + \mathcal{L}_\beta K_{ij}. \tag{2.122}$$

Here we have made the abbreviation

$$M_{ij} = 2S_{ij} - \gamma_{ij}(S - \rho), \tag{2.123}$$

where

$$S_{ij} = \gamma^a_{\ i}\gamma^b_{\ j} T_{ab} \tag{2.124}$$

is the stress according to a normal observer and $S \equiv \gamma^{ij} S_{ij}$ is its trace. The role of the Ricci tensor, determined by γ_{ij} through (2.116), is the same as that of the first term on the right-hand side of (2.93); in particular we notice that both contain different permutations of second derivatives of the "blue" variable.

> **Exercise 2.26** (a) Use the identity
>
> $$\delta \ln \gamma = \gamma^{ij} \delta \gamma_{ij}, \tag{2.125}$$
>
> where δ denotes a general differential operator, to show that the determinant γ of the spatial metric satisfies
>
> $$\partial_t \ln \gamma^{1/2} = -\alpha K + D_i \beta^i. \tag{2.126}$$
>
> (b) Use the definition of the inverse metric γ^{ij}, i.e. $\gamma^{ij}\gamma_{jk} = \delta^i_{\ k}$, to show that
>
> $$\partial_t \gamma^{il} = -\gamma^{ij}\gamma^{lk}\partial_t \gamma_{kj}. \tag{2.127}$$

---

[17] See, e.g., Section 2.6 and 2.7 of Baumgarte and Shapiro (2010) for a derivation.

(c) Use the results of parts (a) and (b) to show that the mean curvature satisfies

$$\partial_t K = -D^2\alpha + \alpha\left(K_{ij}K^{ij} + 4\pi(\rho + S)\right) + \beta^i D_i K. \qquad (2.128)$$

Exercise 2.27 Revisit the Schwarzschild solution in the trumpet coordinate system (1.48) of exercises 1.6 and 2.20, and compute the extrinsic curvature.

(a) Recognize that the solution is static, so that $\partial_t\gamma_{ij} = 0$, then use (2.121) to find K_{ij}. *Hint:* For the purposes of this exercise, it may be easiest to express the Lie derivative in terms of partial derivatives, $\mathcal{L}_\beta\gamma_{ij} = \beta^l\partial_l\gamma_{ij} + \gamma_{lj}\partial_i\beta^l + \gamma_{il}\partial_j\beta^l$. *Check:* One component is $K_{\bar{r}\bar{r}} = -M/\bar{r}^2$.

(b) Check your result from part (a) by computing the mean curvature $K = \gamma^{ij}K_{ij}$ in two different ways: (i) by taking the trace of your result from part (a), and (ii) from equation (2.126), where you may want to use the three-dimensional version of (A.45), $D_i V^i = \gamma^{-1/2}\partial_i(\gamma^{1/2}V^i)$.

The constraint equations (2.117) and (2.119) together with the two evolution equations (2.121) and (2.122) form the famous "ADM" equations, named after Arnowitt, Deser, and Misner.[18] The color coding emphasizes the similarity in the structure of these equations to those for the scalar field of Section 2.1 and those for electrodynamics in Section 2.2 – as we will discuss in more detail in the following section.

First, though, let us recall the discussion at the end of Section 2.2.2, where we reviewed the steps involved in a 3+1 decomposition of Maxwell's equations. We now see that we have pursued very similar steps in general relativity. We first decomposed the fundamental quantity, the spacetime metric g_{ab}. The spatial projection leads to the spatial metric (equation 2.100), which is now one of our spatial dynamical variables. Its time derivative is related to the second dynamical variable, the extrinsic curvature. A purely normal projection of Einstein's equations then leads to the Hamiltonian constraint (2.117), a mixed normal–spatial projection to the momentum constraint (2.119), and a purely spatial projection to the evolution equation (2.122). The initial data now have to satisfy both constraint equations; their time evolution is then determined by the evolution equations. Just as the constraint equation in electrodynamics is preserved by the evolution equations – i.e. data that satisfy the constraint at some initial time will continue to do so when they are evolved with the evolution equations (see exercise 2.1) – the gravitational constraint equations are also preserved by the gravitational evolution equations.[19]

---

[18] See Arnowitt et al. (1962); also York (1979).
[19] See, e.g., Section 2.5 in Alcubierre (2008); Section VIII.7.2 in Choquet-Bruhat (2015); and Section 2.1.6 in Shibata (2016).

Box 2.1 Comparison of 3+1 field equations – version 1

Scalar field

Evolution equations

$$\partial_t \psi = -\kappa,$$ (2.129)

$$\partial_t \kappa = -D^2 \psi + 4\pi \rho.$$ (2.130)

Electrodynamics

Evolution equations

$$\partial_t A_i = -E_i - D_i \phi,$$ (2.131)

$$\partial_t E_i = -D^2 A_i + D_i D^j A_j - 4\pi j_i.$$ (2.132)

Constraint equation

$$D_i E^i = 4\pi \rho \quad \text{(Gauss)}.$$ (2.133)

General relativity

Evolution equations

$$\partial_t \gamma_{ij} = -2\alpha K_{ij} + \mathcal{L}_\beta \gamma_{ij},$$ (2.134)

$$\partial_t K_{ij} = \alpha(R_{ij} - 2K_{ik}K^k_{\ j} + K K_{ij}) - D_i D_j \alpha + 4\pi \alpha M_{ij} + \mathcal{L}_\beta K_{ij}.$$ (2.135)

Constraint equations

$$R + K^2 - K_{ij}K^{ij} = 16\pi\rho \quad \text{(Hamiltonian)},$$ (2.136)

$$D_j(K^{ij} - \gamma^{ij} K) = 8\pi j^i \quad \text{(momentum)}.$$ (2.137)

2.4 Comparison and Outlook

In Section 2.1 we emphasized that the equations for scalar fields are not hard to solve numerically. Now we can compare those equations with the corresponding equations in electrodynamics and general relativity – in fact, we list these equations in Box 2.1 for an easy and direct comparison. In Box 2.2 we list the variables encountered in the three different theories and characterize the associated systems of equations.

In Box 2.1 we write the equations for the scalar field, electrodynamics, and general relativity in the 3+1 forms obtained in Sections 2.1, 2.2.1, and 2.3 respectively. Box 2.2 highlights the hierarchy of these

---

Box 2.2 "Cast" of 3+1 variables

	Scalar field	Electro-dynamics	General relativity	Color
Dynamical variables	ψ	A_i	γ_{ij}	blue
	κ	E_i	K_{ij}	red
Gauge variables	–	ϕ	α, β^i	gold
Matter variables	ρ	ρ, j^i	ρ, j^i, M_{ij}	green
Number of evolution eqns.	2	6	12	
Number of constraint eqns.	0	1	4	
Number of gauge choices	0	1	4	
Dynamical degrees of freedom[a]	1	2	2	

[a] The dynamical degrees of freedom specify the number of independent wave polarization states. For general relativity, there are 12 independent components $(\gamma_{ij}, K_{ij}) - 4$ constraints $- 4$ coordinate (gauge) choices, resulting in 2 independent conjugate pairs (γ_{ij}, K_{ij}), or 2 polarization states for a gravitational wave.

---

three different sets of equations in terms of their gauge freedom: there is none for the scalar field, there is one gauge choice in electrodynamics, expressed by ϕ, and a four-fold gauge freedom in general relativity, expressed by the lapse function α and the shift vector β^i. It is sometimes also useful to incorporate the additional coordinate freedom expressed by α and β^i in the equations for the scalar field and electrodynamics. This results in 3+1 decompositions in the forms shown in the previous sections and listed in Box 2.3. This alternative version of the equations highlights the similarity of the new terms introduced by allowing for this additional coordinate freedom in all three theories.

The color coding helps us to see the similarities between these sets of equations. In all cases the first equation describes the time derivative of the first dynamical variable (the "blue" variable) in terms of the second dynamical variable (the "red" variable), while the second equation describes the time derivative of this second, "red", dynamical variable in terms of second spatial derivatives of the "blue" variable as well as "green" matter sources. In electrodynamics and general relativity we naturally find "gold" gauge terms on the right-hand sides (Box 2.1). Also note that, in general relativity, the spatial derivatives of the "blue" variable are hidden in the Ricci tensor. Evidently, the three different sets of equations have a very similar structure, and therefore one might hope that the same simple numerical methods that work for scalar fields

---

Box 2.3 Comparison of 3+1 field equations – version 2

Scalar field

Evolution equations

$$\partial_t \psi = -\alpha \kappa + \mathcal{L}_\beta \psi, \tag{2.138}$$

$$\partial_t \kappa = -D_a(\alpha D^a \psi) + \alpha \kappa K + 4\pi \alpha \rho + \mathcal{L}_\beta \kappa. \tag{2.139}$$

Electrodynamics

Evolution equations

$$\partial_t A_a^\perp = -\alpha E_a - D_a(\alpha \phi) + \mathcal{L}_\beta A_a^\perp, \tag{2.140}$$

$$\partial_t E^a = D_b(\alpha D^a A_\perp^b - \alpha D^b A_\perp^a) + \alpha E^a K - 4\pi \alpha j_\perp^a + \mathcal{L}_\beta E^a. \tag{2.141}$$

Constraint equation

$$D_i E^i = 4\pi \rho \quad \text{(Gauss)}. \tag{2.142}$$

General relativity

Evolution equations

$$\partial_t \gamma_{ij} = -2\alpha K_{ij} + \mathcal{L}_\beta \gamma_{ij}, \tag{2.143}$$

$$\partial_t K_{ij} = \alpha(R_{ij} - 2K_{ik}K^k_{\ j} + K K_{ij}) - D_i D_j \alpha + 4\pi \alpha M_{ij} + \mathcal{L}_\beta K_{ij}. \tag{2.144}$$

Constraint equations

$$R + K^2 - K_{ij}K^{ij} = 16\pi \rho \quad \text{(Hamiltonian)}, \tag{2.145}$$

$$D_j(K^{ij} - \gamma^{ij} K) = 8\pi j^i \quad \text{(momentum)}. \tag{2.146}$$

---

may also work for electrodynamics and, more importantly from our perspective, for general relativity. This, however, is essentially where the similarities end, and it is time to discuss some of the differences.

For a start, we see that the dynamical variables for the scalar fields are just scalars – as the name suggests – while the variables in electrodynamics and general relativity are rank-1 and rank-2 tensors, respectively (see also Box 2.2). This is not a huge issue, however. It does mean that, in a numerical code, we now need to reserve memory for each component of a tensor, and that we need to evaluate more terms, but this by itself does not pose particularly difficult new issues.

More importantly, we notice that the equations of electrodynamics and general relativity include constraint equations while the scalar fields are unconstrained. For scalar fields we can choose the initial data ψ and κ as arbitrary functions of the spatial coordinates, while in electrodynamics and general relativity the initial data have to satisfy the constraint equations. In electrodynamics, for example, we cannot choose the initial electric field arbitrarily; instead we have to solve Gauss's law (2.133) in order to obtain valid initial data. In general relativity, we similarly have to solve the Hamiltonian constraint (2.136) and the momentum constraint (2.137) in order to find initial data that are consistent with Einstein's equations. This "initial-value" problem will be our focus in Chapter 3.

We also notice that the equations of electrodynamics and general relativity involve gauge variables, while the equations for the scalar field do not.[20] As we have discussed before, the appearance of this gauge freedom comes hand-in-hand with the appearance of constraint equations. Essentially, we start out with equal numbers of equations and variables but, since the constraint equations are redundant with the evolution equations, the variables end up being underdetermined – hence the gauge freedom. In electrodynamics the gauge variable ϕ is related to the constraint equation (2.133), while in general relativity the lapse α and shift vector β^i, which encode the fourfold gauge freedom, are related to the Hamiltonian constraint (2.136) and the momentum constraint (2.137). Recall from our discussion in Section 2.2.3 that the lapse and shift describe the rate at which we wish to advance time and the motion of the spatial coordinates as we move from time slice to time slice. Accordingly, it is not entirely surprising that the lapse and shift make appearances only in the evolution equations, not in the constraint equations which hold on the slice. The field equations do not determine the gauge variables, and we therefore have to make wise choices for these quantities before we can solve the evolution equations – more on this in Section 4.2.

Finally, we notice that the right-hand side of (2.130) for the scalar field involves only the Laplace operator, while both (2.132) in electrodynamics and (2.135) for general relativity involve other spatial second derivatives as well; in electrodynamics we also encounter the gradient of the divergence, and in general relativity we have all the mixed second derivatives hidden in the Ricci tensor; see equations

---

[20] In some sense there is actually a gauge freedom for scalar fields, namely an overall constant, in complete analogy to the freedom to choose an arbitrary reference point for the Newtonian potential. Here, however, we are interested in the freedom to choose arbitrary functions, not just an overall constant.

(1.34) or (2.116). The origin of the Laplace operator in (2.130) was the simple wave equation (2.2), which, as we discussed in Section 2.1, is a symmetric hyperbolic equation and hence has very desirable mathematical properties. Unfortunately, the appearance of the other second-derivative terms in (2.132) and (2.135) destroys these desirable properties. In fact, numerical implementations of the ADM equations typically become unstable and crash after short times. The good news, however, is that we can find reformulations of the equations that avoid these problems, as we will discuss in more detail in Section 4.1.

Exercise 2.28 Repeat for scalar and electromagnetic fields the analysis given at the bottom of Box 2.2 to determine the number of wave polarization states in each.

3

Solving the Constraint Equations

In Chapter 2 we saw that, in a 3+1 decomposition, Einstein's equations split into constraint and evolution equations. The former impose conditions on the gravitational fields at any instant of time, including an initial time, while the latter determine the time evolution of the fields. In this chapter we will discuss some of the conceptual issues one encounters, as well as some strategies one may employ, when solving the constraint equations.

3.1 Freely Specifiable versus Constrained Variables

We will discuss strategies for solving Einstein's constraint equations, i.e. the Hamiltonian constraint (2.136) and the momentum constraint (2.137), but, as before, we will start our discussion with an analogy drawn from electrodynamics. The electrodynamic constraint equation is Gauss's law (2.133),

$$D_i E^i = 4\pi\rho, \tag{3.1}$$

which the electric field E^i has to satisfy at any instant of time. In particular, the electric field cannot take arbitrary values at the initial time – instead it has to be a solution to the constraint equation (3.1). Constructing initial data therefore entails solving the constraint equations. Once solutions to the constraint equations have been found, the fields can be evolved using the evolution equations. According to exercise 2.1 the evolution equations guarantee – at least analytically – that the fields will satisfy the constraints at all times if they satisfy them initially.

How, then, can we solve the constraint equations? We notice a conceptual issue immediately: equation (3.1) forms one equation for the three independent spatial components of the electric field, E^i, leaving the electric field underdetermined. This is not entirely counterintuitive. Consider static fields, for example. For static fields, the constraint equation (3.1) does determine the electric field, because we have the additional information that such a field must the curl-free. Essentially, such "Coulombic", or "longitudinal", fields are completely determined by the charge density ρ present in our space. When allowing for dynamical fields, however, we may add solutions describing electro-magnetic radiation that are not curl-free – they are sometimes called "transverse fields". These fields do not depend on the charge distribution at the current time t (which, if nothing else, would violate causality); instead, they depend on radiation sources at some retarded time in the past. Therefore, these fields cannot be determined from the constraint equation (3.1) alone, and hence the constraint equation can give us only partial information.

We then have to ask what information the constraint equation can give us. One approach would be to solve the constraint equation for one component of the electric field, having made assumptions for the others. For example, we could choose a functional behavior for E^x and E^y, say, and then solve (3.1) for E^z, subject to some boundary condition. This may be possible, but it certainly is not very appealing. For example, for a spherically symmetric charge distribution, the resulting field would end up not being radial for the wrong choices of E^x and E^y. In fact, why would we treat different components of a vector differently? Another approach, one that avoids this asymmetry between different components, might be to choose the direction of the electric field everywhere, and then let the constraint equation determine the magnitude. One might try to implement this latter approach by writing the electric field E^i as

$$E^i = \phi \bar{E}^i, \qquad (3.2)$$

where the $\bar{E}^i$ encode the direction of the electric field (for example, we might choose $\bar{E}^i$ to be unit vectors). We would then hope that the constraint equation would determine the scale factor ϕ, from which we could find the magnitude of the electric field. Evidently, there is freedom as to which information we choose and which information we determine from the constraint.

In reality, there is a better approach to solving the constraint equation (3.1), which we will return to when we discuss solutions to the momentum constraint in Section 3.3.1 below, but the above argument illustrates an important point. In the absence of symmetries, the constraint equations leave the fields underdetermined. Therefore we cannot solve the constraints for all parts of the fields; instead we have to choose some information and solve for the rest. This means that we first have to decide *which* information we want to choose, and which information we want to solve for. Choosing a *decomposition* of the constraint equation helps us to make this choice, i.e. of separating *freely specifiable* variables from *constrained* variables.

3.2 Solving the Hamiltonian Constraint

3.2.1 Conformal Decompositions

Solving the initial-value problem in general relativity entails finding solutions for the spatial metric γ_{ij} and the extrinsic curvature K_{ij} (and possibly expressions for matter sources that enter the stress–energy tensor T^{ij}). The Hamiltonian constraint (2.117)

$$R + K^2 - K_{ij}K^{ij} = 16\pi\rho \qquad (3.3)$$

together with the momentum constraint (2.119)

$$D_j(K^{ij} - \gamma^{ij}K) = 8\pi j^i \qquad (3.4)$$

provide four equations for the twelve components in γ_{ij} and K_{ij}. In this section we will focus on the Hamiltonian constraint (3.3) and will decompose it in such a way that we can constrain the spatial metric γ_{ij}.

To get started, we remind the reader that the Ricci scalar R contains up to second derivatives of the spatial metric. It forms a rather complicated differential operator, however, and *a priori* it is not clear at all how to solve this operator for any part of the metric. To help with this we will introduce a *conformal decomposition*.[1]

Specifically, we will write the spatial metric as

$$\gamma_{ij} = \psi^4 \bar{\gamma}_{ij}. \qquad (3.5)$$

---

[1] This extremely powerful technique was pioneered by Lichnerowicz (1944) and York (1971).

We now refer to $\bar{\gamma}_{ij}$ as the *conformally related metric* and to ψ as the *conformal factor*. The exponent 4 is chosen for convenience, and some authors use different exponents. We sometimes call γ_{ij} the *physical metric* in order to distinguish it from $\bar{\gamma}_{ij}$. Note also the similarity to (3.2) – shortly we will make the components of $\bar{\gamma}_{ij}$ our freely specifiable variables, in analogy to choosing the direction of the electric fields $\bar{E}^i$, and will hope that the Hamiltonian constraint results in a manageable equation for the overall scale, now expressed by the conformal factor ψ.

For our purposes we can think of a conformal decomposition as a mathematical trick, in which we write one unknown as a factor of two unknowns in such a way that it will later make our lives easier – that is the goal. At a deeper level, however, the conformal metric $\bar{\gamma}_{ij}$ also determines a conformal geometry, which we could explore with the same exact tools as we developed in Section 1.2 for the spacetime metric and again in Section 2.3 for the spatial metric. For a starter, we can introduce a conformally related inverse metric $\bar{\gamma}^{ij}$ such that $\bar{\gamma}^{ik}\bar{\gamma}_{kj} = \delta^i{}_j$; this implies that we must have

$$\gamma^{ij} = \psi^{-4}\bar{\gamma}^{ij}. \tag{3.6}$$

Next we define a covariant derivative associated with the conformally related metric. We will do this in complete analogy to (2.113), but with every occurrence of the physical metric replaced by the conformal metric. For example, we write

$$\bar{D}_i V^j = \partial_i V^j + V^k \bar{\Gamma}^j_{ki}, \tag{3.7}$$

where the conformally related Christoffel symbols are given by

$$\bar{\Gamma}^i_{jk} = \frac{1}{2}\bar{\gamma}^{il}(\partial_k \bar{\gamma}_{jl} + \partial_j \bar{\gamma}_{lk} - \partial_l \bar{\gamma}_{jk}) \tag{3.8}$$

(*cf.* 2.114). In fact, we can express the physical Christoffel symbols (2.114) as conformally related Christoffel symbols plus some extra terms; given that the Christoffel symbols involve derivatives of the metric, it is not surprising that these extra terms involve derivatives of the conformal factor.

Exercise 3.1 Show that

$$\Gamma^i_{jk} = \bar{\Gamma}^i_{jk} + 2\left(\delta^i{}_j \partial_k \ln \psi + \delta^i{}_k \partial_j \ln \psi - \bar{\gamma}_{jk}\bar{\gamma}^{il}\partial_l \ln \psi\right). \tag{3.9}$$

Continuing to follow in our footsteps in Section 1.2 we next define conformal versions of the Riemann tensor, Ricci tensor, and Ricci scalar. As in (3.9) we can relate these to their physical counterparts by adding extra terms: we insert (3.5) into the expressions for these tensors and use the product rule to separate derivatives of $\bar{\gamma}_{ij}$ from those of ψ, whereby the former then yield conformal versions of the tensors while the latter give us the extra terms. Given that the Riemann and Ricci tensors contain second derivatives of the metric, the extra terms now include second derivatives of the conformal factor. We will skip details of this derivation[2] and will instead get straight to the key point. Relating the physical Ricci scalar R to its conformal counterpart $\bar{R}$ we find[3]

$$R = \psi^{-4}\bar{R} - 8\psi^{-5}\bar{D}^2\psi, \tag{3.10}$$

where we have defined the "conformal Laplace operator"

$$\bar{D}^2\psi \equiv \bar{\gamma}^{ij}\bar{D}_i\bar{D}_j\psi. \tag{3.11}$$

Inserting (3.10) into the Hamiltonian constraint we now obtain

$$\bar{D}^2\psi - \frac{\psi}{8}\bar{R} + \frac{\psi^5}{8}\left(K_{ij}K^{ij} - K^2\right) = -2\pi\psi^5\rho. \tag{3.12}$$

This is nice! We now declare the conformally related metric components $\bar{\gamma}_{ij}$ to be our freely specifiable variables. Making a choice for $\bar{\gamma}_{ij}$ then determines $\bar{R}$ as well as for the coefficients in the Laplace operator (3.11). As long as $\bar{\gamma}_{ij}$ is sufficiently well behaved, this operator can be inverted. Given values for the extrinsic curvature K_{ij} as well as for the matter source density ρ, the Hamiltonian constraint (3.12) then forms a Poisson-type equation for the conformal factor ψ. In general this equation can be solved with standard numerical techniques – see Appendix B for an example – but we can also see this formalism at work for some simple yet important analytical solutions.

3.2.2 Schwarzschild and Brill–Lindquist Solutions

In order to obtain analytical solutions to the Hamiltonian constraint (3.12) we have to make simple choices for the matter sources ρ, the extrinsic curvature K_{ij}, and the conformally related metric $\bar{\gamma}_{ij}$. Specifically, we will focus on *vacuum* spacetimes, so that $\rho = 0$ as

---

[2] See, e.g., Section 3.1 in Baumgarte and Shapiro (2010) for more details.
[3] We assume here that ψ can be treated as a scalar regarding its covariant derivatives.

well as $j^i = 0$ in the momentum constraint (3.4). We will also assume a moment of *time symmetry*, whereby all time derivatives of γ_{ij} are zero and the four-dimensional line interval has to be invariant under time reversal, $t \rightarrow -t$. The latter condition implies that the shift must satisfy $\beta^i = 0$ and hence, by equation (2.134), the extrinsic curvature must vanish everywhere on the slice: $K_{ij} = 0 = K$. With these choices, the momentum constraint (3.4) is satisfied identically. Finally, we choose *conformal flatness*, meaning that we choose the conformally related metric to be flat,

$$\bar{\gamma}_{ij} = \eta_{ij}. \tag{3.13}$$

With this latter choice, the conformally related Ricci scalar $\bar{R}$ in (3.12) vanishes (the curvature of flat space is zero!), and the Laplace operator (3.11) becomes the common-or-garden-variety flat-space Laplace operator. That means that the Hamiltonian constraint (3.12) reduces to the remarkably simple Laplace equation

$$\bar{D}^2 \psi = 0. \tag{3.14}$$

The specific form of the Laplace operator depends on the coordinates chosen. In spherical symmetry, for example, equation (3.14) becomes

$$\frac{1}{r^2} \frac{d}{dr} \left(r^2 \frac{d\psi}{dr} \right) = 0 \tag{3.15}$$

(see exercise A.9), which is solved by

$$\psi = A + \frac{B}{r}, \tag{3.16}$$

where A and B are two arbitrary constants. We will choose $A = 1$, so that asymptotically, as $r \rightarrow \infty$, the conformal factor approaches unity and the physical metric (3.5) becomes the flat metric – we then say that our metric is "asymptotically flat". We also rename B as $M/2$, so that the conformal factor becomes

$$\psi = 1 + \frac{M}{2r}. \tag{3.17}$$

This, in fact, is an old friend, who we previously met in exercises 1.5 and 2.19; the spatial metric $\gamma_{ij} = \psi^4 \eta_{ij}$ is the metric of a Schwarzschild black hole on a slice of constant Schwarzschild time, expressed in isotropic coordinates. We could analyze the properties of this metric to find that the constant M corresponds to the gravitational mass of the black hole, i.e. the mass that an observer at a large separation would measure.[4]

Exercise 3.2 Revisit the Schwarzschild solution in the trumpet coordinate system (1.48) of exercises 1.6, 2.20, and 2.27. Presumably you found $\gamma_{ij} = \psi^4 \eta_{ij}$ with $\psi = (1 + M/\bar{r})^{1/2}$ in exercise 2.20, meaning that we can identify $\bar{\gamma}_{ij} = \eta_{ij}$ and use the simplifications of conformal flatness. Use your result from exercise 2.27 to compute K^2 and $K_{ij}K^{ij}$, then compute the Laplace operator $\bar{D}^2\psi$, and verify that the solution satisfies the Hamiltonian constraint (3.12) with $\rho = 0$.

So far, so good, even though a skeptic could argue that we have just spent a lot of effort on rederiving a solution that we already knew. Here is something neat, though: under the assumptions that we made above, the Hamiltonian constraint (3.14) has become a linear equation. That means that we can use the superposition principle to construct initial data describing multiple black holes. For N black holes, say, with masses M_n and centered on coordinate points x^i_n, we obtain the famous *Brill–Lindquist* initial data[5]

$$\psi = 1 + \sum_{n=1}^{N} \frac{M_n}{2r_n}, \tag{3.18}$$

where $r_n \equiv |x^i - x^i_n|$ measures the coordinate distance at a coordinate location x^i from the nth black hole at x^i_n.

While equation (3.18) provides an exact solution for multiple black holes, which is quite remarkable, we should caution that this solution holds only at the initial moment of time symmetry, of course. In order to find the solution at all times we would have to solve the evolution equations, as we will discuss in Chapter 4. Those equations are no longer linear, and we cannot simply use the superposition principle. Similarly, the Schwarzschild metric (2.109) is a spacetime solution that is static at *all* times. It also satisfies nonlinear equations, so that we cannot use the superposition principle to construct complete spacetime solutions for multiple black holes. We were fortunate, however, that under the assumptions made above the Hamiltonian constraint becomes linear, allowing us to construct initial data for multiple black holes.

---

[4] For example, it is the mass that would give the correct period of a test particle orbiting at large distances according to Kepler's third law.
[5] See Brill and Lindquist (1963).

Another limitation of the above solution describing black holes at a moment of time symmetry is that it is restricted to an instant at which the black holes are momentarily at rest. To construct initial data describing two black holes in a binary orbit we therefore need to abandon the assumption of time symmetry, meaning that we have to allow for a non-zero extrinsic curvature – and that means that we have to solve the momentum constraint together with the Hamiltonian constraint.

3.3 Solving the Momentum Constraints

3.3.1 The Conformal Transverse-Traceless Decomposition

The Hamiltonian constraint (3.12) for the conformal factor ψ is a nonlinear version of the Poisson equation. For static electric fields we may write $E_i = -D_i\phi$, insert this into the constraint (3.1), and obtain a similar Poisson equation for the electrostatic potential ϕ. In general, however, the momentum constraint (3.4) more closely resembles the electromagnetic constraint in the form (3.1), since, in the color coding of Chapter 2, both involve the divergence of red (dynamical) variables. Accordingly, we can use techniques to solve the momentum constraint that are quite similar to those that may be familiar from electrodynamics.

Recall from electrodynamics the Helmholtz theorem, which states that we can write any vector as a sum of two parts, one of which has zero curl and the other has zero divergence. We call the former the longitudinal (L) part and the latter the transverse (T) part. Since the curl of a gradient vanishes identically, we can also write the longitudinal part as the gradient of some function. Applying the above to the electric field we have

$$E^i = E_L^i + E_T^i = -D^i\phi + E_T^i, \qquad (3.19)$$

where we have set $E_L^i = -D^i\phi$. We could also write E_T^i as the curl of a vector potential, but that would be less relevant to our discussion – all that matters for our purposes here is that $D_i E_T^i = 0$. Inserting (3.19) into (3.1) we obtain

$$D_i E^i = -D_i D^i\phi + D_i E_T^i = -D^2\phi = 4\pi\rho, \qquad (3.20)$$

a nice Poisson-type equation for the potential ϕ. Note that this allows for a clean separation of the freely specifiable and the constrained variables:

we can freely choose E_T^i (amounting to two independent variables, since E_T^i has to be divergence-free), while E_L^i (determined by one independent variable, namely ϕ) is evidently constrained.

We will consider a similar decomposition of the extrinsic curvature K_{ij} to solve the momentum constraint

$$D_j(K^{ij} - \gamma^{ij} K) = 8\pi j^i. \tag{3.21}$$

First, though, we split K_{ij} into its trace, $K = \gamma^{ij} K_{ij}$, and a traceless part A_{ij} with, as the name suggests, zero trace, so that $\gamma^{ij} A_{ij} = 0$:

$$K_{ij} = A_{ij} + \frac{1}{3}\gamma_{ij}K. \tag{3.22}$$

Recall that the trace K is often referred to as the *mean curvature*. We now treat A_{ij} and K as independent functions, even though both are related to K_{ij}, of course. In fact, we can always reassemble K_{ij} from (3.22). As was the case for the spatial metric in Section 3.2.1, it is again useful to consider a conformal decomposition for A_{ij} and K. Since K is a scalar, its derivative in (3.21) is just a partial derivative, and there is no particular advantage in conformal rescaling – we will therefore leave it as it is. The traceless part A^{ij}, however, will appear in the divergence $D_i K^{ij}$. Expanding this divergence involves Christoffel symbols. Expressing these Christoffel symbols in terms of their conformally related counterparts, as in (3.9), introduces derivatives of the conformal factor ψ. It turns out that these derivative terms cancel when we transform the divergence conformally, leaving us with

$$D_j A^{ij} = \psi^{-10}\bar{D}_j \bar{A}^{ij}, \tag{3.23}$$

if we transform A_{ij} according to

$$A^{ij} = \psi^{-10}\bar{A}^{ij}. \tag{3.24}$$

Exercise 3.3 Insert (3.9) and (3.24) into the covariant derivative associated with the physical spatial metric,

$$D_i A^{ij} = \partial_i A^{ij} + A^{kj}\Gamma^i_{jk} + A^{ik}\Gamma^j_{jk}, \tag{3.25}$$

and then identify the corresponding covariant derivative with the conformally related metric,

$$\bar{D}_i A^{ij} = \partial_i A^{ij} + A^{kj}\bar{\Gamma}^i_{jk} + A^{ik}\bar{\Gamma}^j_{jk}, \tag{3.26}$$

to derive (3.23).

We lower the indices on $\bar{A}^{ij}$ with $\bar{\gamma}_{ij}$; this implies that

$$A_{ij} = \psi^{-2}\bar{A}_{ij}. \tag{3.27}$$

Now we are ready to decompose $\bar{A}_{ij}$ in a way similar to (3.19). As a generalization of the Helmholtz theorem, we can write a traceless rank-2 tensor as a sum of a longitudinal part that can be written as the *vector gradient* of a vector W^i,

$$\bar{A}_{\mathrm{L}}^{ij} = (\bar{L}W)^{ij} \equiv \bar{D}^i W^j + \bar{D}^j W^i - \frac{2}{3}\bar{\gamma}^{ij}\bar{D}_k W^k, \tag{3.28}$$

and a transverse-traceless part whose divergence vanishes,

$$\bar{D}_j \bar{A}_{\mathrm{TT}}^{ij} = 0. \tag{3.29}$$

The divergence of $\bar{A}^{ij}$ therefore becomes

$$\bar{D}_j \bar{A}^{ij} = \bar{D}_j \bar{A}_{\mathrm{L}}^{ij} = \bar{D}_j (\bar{L}W)^{ij}$$

$$= \bar{D}_j \bar{D}^j W^i + \frac{1}{3}\bar{D}^i(\bar{D}_j W^j) + \bar{R}^i{}_j W^j \equiv (\bar{\Delta}_L W)^i, \tag{3.30}$$

where we have used the three-dimensional analogue of equation (D.10), contracted on one pair of indices, to commute the second derivatives of W^i, and where we have defined the *vector Laplacian* $\bar{\Delta}_L$.

Just as we inserted (3.19) into Gauss's law to obtain the Poisson-type equation (3.20) for the potential Φ, we now insert

$$K^{ij} = \psi^{-10}\bar{A}_{\mathrm{L}}^{ij} + \psi^{-10}\bar{A}_{\mathrm{TT}}^{ij} + \frac{1}{3}\gamma^{ij}K$$

$$= \psi^{-10}(\bar{L}W)^{ij} + \psi^{-10}\bar{A}_{\mathrm{TT}}^{ij} + \frac{1}{3}\gamma^{ij}K \tag{3.31}$$

into the momentum constraint (3.21), use (3.23) as well as (3.30), and obtain a vector Poisson-type equation for the vector potential W^i,

$$(\bar{\Delta}_L W)^i - \frac{2}{3}\psi^6\bar{\gamma}^{ij}\bar{D}_j K = 8\pi\psi^{10}j^i. \tag{3.32}$$

Exercise 3.4 Confirm equation (3.32).

As in our electrodynamic analogue, this equation suggests a decomposition into freely specifiable and constrained variables in this so-called "conformal transverse-traceless" decomposition. Freely choosing the components of the transverse components $\bar{A}_{\mathrm{TT}}^{ij}$ as well as the mean curvature K, the momentum constraint then constrains the components of the longitudinal part $\bar{A}_{\mathrm{L}}^{ij}$.

It is also useful to account for the six degrees of freedom in K_{ij} (a symmetric rank-2 tensor in three dimensions) in this decomposition. Evidently there is one independent component in the scalar K. It takes a little more thought to realize that there are only two independent components in $\bar{A}_{\mathrm{TT}}^{ij}$: we start with six independent components in a symmetric rank-2 tensor in three dimensions, we lose one because $\bar{A}_{\mathrm{TT}}^{ij}$ is traceless, and we lose three more because its divergence vanishes (see 3.29). The remaining three degrees of freedom in K_{ij} are accounted for by the longitudinal part $\bar{A}_{\mathrm{L}}^{ij}$ and are expressed by the three components of the vector potential W^i.

Combining the transverse-traceless decomposition of the extrinsic curvature with the conformal decomposition of the metric in Section 3.2 completes one possible decomposition of the initial-value problem. In this approach, we can freely choose the conformally related metric $\bar{\gamma}_{ij}$, the mean curvature K, and the transverse-traceless part $\bar{A}_{\mathrm{TT}}^{ij}$ of the extrinsic curvature, and we then solve the Hamiltonian constraint for the conformal factor ψ and the momentum constraint for the vector potential W^i, i.e. the longitudinal part of the extrinsic curvature. There are other useful decompositions of the constraint equations; the *conformal thin-sandwich* decomposition, in particular, provides a natural framework for constructing quasi-equilibrium data, especially in the presence of fluids.[6] One neat aspect of the transverse-traceless decomposition, though, is the existence of analytical solutions for the extrinsic curvature that describe black holes with linear or angular momentum, as we will review in the following section.

3.3.2 Bowen–York Solutions

Before we can construct analytical solutions to the momentum constraint (3.32) we again have to make choices. In order to construct black-hole solutions, we will focus on the vacuum case, which implies that $j^i = 0$ on the right-hand side of (3.32). As in Section 3.2.2 we will choose conformal flatness, i.e. $\bar{\gamma}_{ij} = \eta_{ij}$, so that $\bar{R}^i{}_j = 0$ in (3.32) and all covariant derivatives become flat covariant derivatives or simply partial derivatives, in Cartesian coordinates. Next we have to

---

[6] For textbook treatments see Section 3.3 in Alcubierre (2008); Section 3.3 Baumgarte and Shapiro (2010); Section 9.3 in Gourgoulhon (2012); Section 5.6 in Shibata (2016).

make choices for the freely specifiable parts of the extrinsic curvature. We will choose the transverse part to vanish, $\bar{A}^{ij}_{\mathrm{TT}} = 0$, but that will play a role only later since $\bar{A}^{ij}_{\mathrm{TT}}$ does not even make an appearance in (3.32). Of more immediate importance is our choice for the mean curvature. We will choose *maximal slicing*, by which we mean

$$K = 0. \tag{3.33}$$

The mean curvature K vanishes for spatial slices that, given some boundary conditions, maximize the volume of the slice as measured by normal observers – hence the name. A familiar analogy is provided by soap films suspended from a (possibly bent) closed hoop, which we can think of as two-dimensional slices of a three-dimensional Euclidean space. As long as we can neglect gravity, the surface tension of the soap minimizes the surface area of the film. For the surface of such films we can also define normal vectors and the extrinsic curvature as well as its trace, the mean curvature K, and we would find that the latter vanishes. The (three-dimensional) spatial slices that we are considering here are embedded in a (four-dimensional) Lorentzian geometry rather than a Euclidean geometry, though; imposing $K = 0$ therefore maximizes the volume rather than minimizing the area.

Choosing maximal slicing has an important effect in (3.32), because now all terms involving the conformal factor ψ vanish. That means that the momentum constraint decouples from the Hamiltonian constraint, and we can solve the former independently of the latter.

The momentum constraint now takes the form of a vector Laplace equation, and, in Cartesian coordinates, can be written as

$$\partial^j \partial_j W^i + \frac{1}{3} \partial^i \partial_j W^j = 0. \tag{3.34}$$

Note that this equation is already linear so that, once we have a solution for one black hole, we will be able to use the superposition principle to construct solutions for multiple black holes. There are tricks that can be used to solve equations involving the vector Laplacian,[7] but rather than actually deriving the Bowen–York solutions we will just quote and verify them here.

Consider a black hole centered on a coordinate location x^i_{BH}. At a coordinate location x^i we then define the (coordinate) unit vector pointing away from the black hole as

---

[7] See, e.g., Appendix B in Baumgarte and Shapiro (2010).

$$l^i = \frac{x^i - x^i_{BH}}{s},$$ (3.35)

where $s = \left((x_i - x^{BH}_i)(x^i - x^i_{BH})\right)^{1/2}$ is the coordinate distance between the two locations. It is useful to verify that $\partial_i s = l_i$ and

$$\partial_i l_j = \frac{1}{s}\left(\delta_{ij} - l_i l_j\right).$$ (3.36)

We now claim that

$$W^i = \frac{\bar{\epsilon}^{ilk} l_l J_k}{s^2}$$ (3.37)

is a solution to (3.34). Here $\bar{\epsilon}^{ilk}$ is the three-dimensional and conformally related version of the completely antisymmetric Levi–Civita symbol, which we encountered in Section 2.2.2; in our present context it is +1 for an even permutation of xyz, −1 for an odd permutation, and 0 otherwise. Also, J_k, not to be confused with the momentum density j^i, is taken to be a vector whose components are constant when expressed in Cartesian coordinates – we will discuss the physical interpretation of this vector shortly.

First, though, we will verify our claim. From (3.37) we compute the first and second derivatives, obtaining

$$\partial_j W^i = \frac{\bar{\epsilon}^{ilk} J_k}{s^3}(\delta_{jl} - 3l_j l_l)$$ (3.38)

and

$$\partial^m \partial_j W^i = \frac{3\bar{\epsilon}^{ilk} J_k}{s^4}\left(5l^m l_j l_l - l^m \delta_{jl} - l_l \delta^m_j - l_j \delta^m_l\right).$$ (3.39)

Setting $i = j$ in the first derivative (3.38) results in the contraction of an antisymmetric tensor with a symmetric tensor (see exercise A.2); we must therefore have $\partial_j W^j = 0$, and the second term in (3.34) vanishes already. To compute the first term in (3.34) we set $m = j$ in the second derivative of (3.39); recognizing that $l^j l_j = 1$ and $\delta^j_j = 3$ we see that this term vanishes also, meaning that (3.37) is indeed a solution to (3.34).

For conformal flatness and in Cartesian coordinates, equation (3.28) reduces to

$$\bar{A}^{ij}_L = \partial^i W^j + \partial^j W^i - \frac{2}{3}\eta^{ij}\partial_k W^k,$$ (3.40)

where again the last term vanishes. Inserting (3.38) we then obtain

$$\bar{A}_{L}^{ij} = \frac{6}{s^3} l^{(i} \bar{\epsilon}^{j)kl} J_k l_l, \tag{3.41}$$

where we have used the symmetrization operator ( ) defined in (A.15).

Exercise 3.5 (a) Consider the vector potential

$$W^i = -\frac{1}{4s}\left(7P^i + l^i l_k P^k\right), \tag{3.42}$$

where the components of the vector P^i are constant when expressed in Cartesian coordinates. Verify that (3.42) also satisfies (3.34). *Check:* You should find that $\partial_i W^i = 3l_i P^i/(2s^2)$.

(b) Show that the extrinsic curvature corresponding to the vector potential (3.42) is

$$\bar{A}_{L}^{ij} = \frac{3}{2s^2}\left(P^i l^j + P^j l^i - (\eta^{ij} - l^i l^j)l_k P^k\right). \tag{3.43}$$

Equations (3.41) and (3.43) form the famous *Bowen–York* solutions.[8] To obtain the extrinsic curvature, we can insert (3.41) or (3.43) into (3.31) and find

$$K^{ij} = \psi^{-10}\bar{A}_{L}^{ij}, \tag{3.44}$$

where our choice $\bar{A}_{TT}^{ij} = 0$, together with the maximal slicing $K = 0$, has finally made an appearance.

Of course, we don't know yet what ψ is – more on that in Section 3.4 below. In the meantime, though, we could evaluate the extrinsic curvature (3.44) at large distances from the black hole where, assuming asymptotic flatness, the conformal factor approaches unity. We could then evaluate the physical properties of the solution (3.41) in that regime. Since this requires some integrals that exceed the scope of this volume,[9] we will just report that the extrinsic curvature (3.41) carries an angular momentum J^i, while the solution (3.43) carries a linear momentum P^i – we named the constant vectors in (3.37) and (3.42) with a little foresight.

Recall that the momentum constraint in the form (3.34) is linear. We can therefore add together several Bowen–York solutions (3.41) and (3.43), and, remarkably, the sum will still be a solution to the momentum constraint. That means that we can simply use superposition to construct

---

[8] See Bowen and York (1980).
[9] See, e.g., Section 4.3 in Poisson (2004); Appendix A in Alcubierre (2008); Section 3.5 of Baumgarte and Shapiro (2010); Section 5.3 in Shibata (2016).

multiple black holes that carry both linear and angular momentum. Likewise, we can construct binary black-hole solutions by adding two Bowen–York solutions centered on two different coordinate locations, with their respective linear momenta P^i corresponding to the orbital velocities at the given binary separation. This, in fact, is a common approach for constructing binary black-hole initial data. We remind the reader, however, that these solutions are solutions to the momentum constraint only. In order to complete the specification of the initial data we also need the conformal factor ψ, which requires revisiting the Hamiltonian constraint.

3.4 Puncture Initial Data for Black Holes

Recall that in Section 3.2.2 we derived some elementary vacuum solutions to the Hamiltonian constraint (3.12) under the assumption of conformal flatness and time symmetry, by which we assumed the extrinsic curvature to vanish. In particular we recovered the Schwarzschild solution (3.17). Now we will relax the assumption of time symmetry, and will instead assume that the extrinsic curvature is given by a Bowen–York solution. Under the assumptions of Section 3.3.2, the Hamiltonian constraint (3.12) then becomes

$$\bar{D}^2\psi + \frac{1}{8}\psi^{-7}\bar{A}^{\mathrm{L}}_{ij}\bar{A}^{ij}_{\mathrm{L}} = 0, \tag{3.45}$$

where we assume $\bar{A}^{ij}_{\mathrm{L}}$ to be given by a Bowen–York solution (3.41) or (3.43), or a superposition of both.

Equation (3.45) is significantly harder than the time-symmetric version (3.14), because it is nonlinear. Given (3.17) we expect that the conformal factor will diverge at the centers of any black holes; we note that the Bowen–York solutions (3.41) or (3.43) also diverge at those centers. Dealing with black-hole singularities is one of the key challenges of numerical relativity – here we encounter a first example.

Several different approaches have been suggested for dealing with the singularities in equation (3.45),[10] but a particularly elegant solution is provided by the so-called *puncture method*.[11]

---

[10] The first successful solution invoked *conformal imaging*, which results in boundary conditions that can be used to eliminate the black-hole interiors; see Cook (1991, 1994). It also turns out that singularities can be avoided altogether by considering an inverse power of the conformal factor; see Baumgarte (2012).

[11] See Brandt and Brügmann (1997); Beig and Ó Murchadha (1994, 1996).

The first key observation in the puncture method is this: close to each black hole's center, which we now refer to as a *puncture*, the conformal factor is dominated by a singular term $\psi \simeq \mathcal{M}/s$, where $\mathcal{M}$ is called the *puncture mass*. We will skip a formal proof but will point out that this is certainly consistent with (3.45). For Bowen–York solutions, the term, $\bar{A}_{ij}^{\mathrm{L}} \bar{A}_{\mathrm{L}}^{ij}$ in (3.45) diverges at most as fast as s^{-6}, but this is then suppressed by the term $\psi^{-7} \propto s^{7}$. Close to the puncture, the Laplace operator acting on the conformal factor ψ must therefore vanish, which, as we have seen in Section 3.2.2, leads to solutions that diverge as $\mathcal{M}/s$.

This observation leads to the following idea. If we already know the singular behavior close to the punctures analytically, let's write the solution ψ as a sum of two terms – one that absorbs these singular terms analytically and a correction term that we need to determine numerically but that we can assume to be regular. Quite generally, it is a good idea to take advantage of as much analytical information as is available in any numerical work. Specifically, we will write

$$\psi = 1 + \frac{1}{\alpha} + u, \qquad (3.46)$$

where

$$\frac{1}{\alpha} = \sum_n \frac{\mathcal{M}_n}{2 s_n} \qquad (3.47)$$

describes the singular behavior $\mathcal{M}_n/s_n$ for each of the n black holes that we are considering, and where u is the correction term. Note that α approaches zero at each puncture $s_n = 0$. Inserting (3.46) into (3.45), and using $\bar{D}^2(1/s_n) = 0$ for all $s_n > 0$, we now obtain an equation for u,

$$\bar{D}^2 u = -\beta \, (\alpha + \alpha u + 1)^{-7}, \qquad (3.48)$$

where we have

$$\beta \equiv \frac{1}{8} \alpha^7 \bar{A}_{ij}^{\mathrm{L}} \bar{A}_{\mathrm{L}}^{ij}. \qquad (3.49)$$

Given the choices for the Bowen–York solutions (3.41) or (3.43), which specify the black holes' locations and their linear and angular momenta, as well as choices for the black holes' puncture masses $\mathcal{M}_n$, the function β becomes an analytical function of the spatial coordinates.

Equation (3.48) is still a nonlinear elliptic equation (we will deal with that in Appendix B.2.2), but now all terms are regular everywhere,

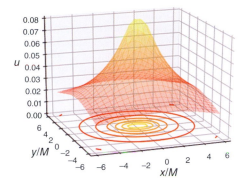

Figure 3.1 The function u, as defined in (3.46), for a black hole with linear momentum $P^i/\mathcal{M} = (1,0,0)$ at the origin. We show data close to the $z = 0$ plane, together with their projected contours for a very coarse resolution of 24^3 grid points, produced with the Python script `puncture.py` provided in Appendix B.2.3.

and the equation allows non-trivial solutions for u that remain regular everywhere. The equation can therefore be solved with standard numerical techniques – in fact, we will provide a Python script `puncture.py` that does just that in Appendix B.2.3. This script serves as a "barebones" example of an initial-data code and allows the reader to explore some numerical techniques that are often employed in these types of problems. An example solution produced with this script is shown in Fig. 3.1. Inserting the solutions u into (3.46) then yields the conformal factor ψ, which completes our construction of black-hole initial data.

To construct binary black-hole initial data, we choose for $\bar{A}_{\mathrm{L}}^{ij}$ a superposition of two Bowen–York solutions separated by a certain binary separation s, one carrying a linear momentum P^i and the other carrying $-P^i$. To endow the individual black holes with spin, we add Bowen–York solutions that carry angular momentum. The value of P^i that corresponds to a circular orbit for a given value of s (or a value that minimizes the eccentricity during the subsequent inspiral) can be found either from energy arguments or by using post-Newtonian expressions. Sequences of initial data which mimic evolutionary sequences in the "adiabatic" regime can then be constructed by varying the binary separation s.[12] Many current binary black-hole simulations adopt puncture initial data as described here.[13]

---

[12] See, e.g., Cook (1994); Baumgarte (2000); Ansorg et al. (2004).
[13] In the Einstein Toolkit, for example, the `TwoPunctures` thorn adopts the algorithm of Ansorg et al. (2004) to construct puncture initial data; see www.einsteintoolkit.org. A spectral

Exercise 3.6   (a) Run the Python script `puncture.py` described in Appendix B.2.3 and available at www.cambridge.org/NRStartingfromScratch to construct puncture initial data for the single black hole shown in Fig. 3.1. Plot the data using `puncture_plot.py`, also available at www .cambridge.org/NRStartingfromScratch, to reproduce the figure.

(b) Read Appendices B.1 and B.2 for an overview of the numerical techniques employed in constructing `puncture.py`.

(c) Identify in `puncture.py` the lines of code that implement the equations in Appendices B.1 and B.2 used to construct the initial data.

We encourage readers to work through exercises B.3–B.5 in Appendix B.2.3 to explore further some numerical techniques involved in solving constraint equations.

3.5 Discussion

We have discussed only one specific approach to constructing black-hole initial data, and, while a detailed presentation of other approaches would go beyond the scope of this volume, we should at least briefly mention some related subjects and issues.

We have already alluded to alternative decompositions of the constraint equations, specifically the conformal thin-sandwich decomposition.[14] This decomposition is particularly appealing for the construction of quasi-equilibrium data. It is commonly used for initial data involving matter, e.g. binary neutron stars in quasi-circular orbits, but it is also being used for black holes.[15]

One limitation of the Bowen–York data of Section 3.3.2 is related to spin. If one sets up Bowen–York solutions with a large angular momentum as initial data, these solutions will settle down to black holes with smaller angular momenta in dynamical evolutions. This phenomenon is probably related to the assumption of conformal flatness in the construction of Bowen–York data. It appears very unlikely that stationary rotating black holes, uniquely described by the Kerr solution, admit conformally flat spatial slices,[16] meaning that spinning Bowen–York black holes are not Kerr black holes and thus they are not in

---

interpolation routine for the `TwoPunctures` thorn that greatly improves the performance is provided by Paschalidis et al. (2013).

[14] See York (1999); Isenberg (1978); Wilson and Mathews (1995) for related earlier versions.

[15] See Gourgoulhon et al. (2002); Grandclément et al. (2002); Cook and Pfeiffer (2004); Caudill et al. (2006). See also Baumgarte and Shapiro (2010).

[16] See Garat and Price (2000); de Felice et al. (2019); see also pp. 72–73 in Baumgarte and Shapiro (2010) for a discussion.

equilibrium. This observation also suggests a solution to the problem, though, which is to adopt spatial slices of the Kerr metric as the conformally related metric, rather than a flat metric.[17]

While we have completed at least one approach to constructing solutions to Einstein's constraint equations, we recognize that we do not yet have any tools to evaluate the physical properties of these solutions. For a start, we would like to know the total gravitational mass[18] as well as the angular and linear momenta of these solutions. Evaluating these requires performing certain integrals over a closed surface surrounding, and at large separations from, the black holes, as we have mentioned before. We would also like to know about the location and surfaces of the black holes in these solutions, which requires locating black-hole *horizons*. As we discussed in Section 1.3.1, we distinguish between event horizons and apparent horizons. Locating the former requires knowledge of the entire spacetime but apparent horizons, when they exist, can be identified from the data on one time slice only. Developing any of these tools, however, goes beyond the scope of this volume. We refer the interested reader to discussions in more detailed textbooks on numerical relativity[19] and will now proceed to the dynamical evolution of initial data.

---

[17] See, e.g., Lovelace et al. (2008) for an implementation of this choice in the context of the conformal thin-sandwich approach, and Ruchlin et al. (2017) in the context of the puncture method.

[18] Spoiler alert: for multiple black holes, the sum of the "puncture mass" parameters $\mathcal{M}_n$ is, in general, *not* the total gravitational mass M.

[19] See, e.g., Sections 6.7 and 6.8 in Alcubierre (2008) or Chapter 7 in either Baumgarte and Shapiro (2010) or Shibata (2016).

4

Solving the Evolution Equations

In Chapter 3 we discussed techniques for solving the constraint equations, resulting in initial data for the spatial metric and extrinsic curvature on a time slice. Now we will assume that we are given such initial data and want to find out how the gravitational fields evolve as time advances. This entails solving the evolution equations.

4.1 Reformulating the Evolution Equations

We start this chapter with a great sense of optimism. In Chapter 2 we kept on emphasizing the similarities between the equations of general relativity and those of scalar fields and electromagnetism. If we know how to solve the evolution equations for the latter, why not for the former? Moreover, in Chapter 3 we learned how to construct initial data – so, do we not now have everything we need to integrate the evolution equations? Not quite, because we saw in Chapter 2 that we need to make choices for the lapse function α and the shift vector β^i in order to determine the dynamical evolution of the coordinates. We will return to that in Section 4.2 below, but for now assume that we have made some reasonable choices, whatever that may mean. Unfortunately, however, there is some bad news: generically, even for suitable coordinate choices, numerical implementations of the ADM equations as presented in Section 2.3 lead to numerical instabilities and will crash in many applications in 3+1 dimensions.

A priori, of course, it is difficult to decide why a particular code becomes unstable and crashes. The instability could be caused by a number of different possibilities. For a starter, the algorithm could have a bug, i.e. there might be an error in it. Perhaps the coordinate conditions are not as suitable as we thought. Perhaps the outer boundary conditions are incorrect. Perhaps we are using a numerical scheme that does not work in this case. Significant effort has gone into exploring all these possibilities, but in the end all the evidence points to the ADM equations themselves.

That is the bad news. The good news is that it is possible to *reformulate* the evolution equations in such a way that they can be solved numerically without the codes crashing. Essentially, two different ingredients are commonly used in such reformulations: (i) we can introduce new variables that absorb some derivatives of already existing variables, and (ii) we can add multiples of the constraint equations to the evolution equations. Mathematicians would explain that these manipulations change the principal part of the differential operator and thus the structure of the highest derivatives, which therefore changes the character of the equation – one formulation may be symmetric hyperbolic, and hence well-posed, while another may not.

Heuristically, however, it is perhaps counterintuitive that the properties of one formulation should be better or worse than that of another. After all, an exact solution to one formulation should be an exact solution to any other formulation – so why is there a difference at all? The answer lies in recognizing that, in this context, we should be less concerned with the exact solution and instead think about errors. We encounter numerical error in (almost) any numerical calculation, and the behavior of this error determines whether an implementation is stable. It turns out that, while exact solutions are the same in all formulations, the error can behave very differently. In order to see this explicitly with just a few lines of calculation we return, as in previous chapters, to electrodynamics.

4.1.1 Maxwell's Equations

In Section 2.2.1 we wrote Maxwell's equations as the pair of evolution equations

$$\partial_t A_i = -E_i - D_i \phi, \tag{4.1}$$

$$\partial_t E_i = -D_j D^j A_i + D_i D^j A_j - 4\pi j_i, \tag{4.2}$$

subject to the constraint equation

$$D_i E^i = 4\pi \rho. \tag{4.3}$$

We now define the *constraint violation* $\mathcal{C}$ as

$$\mathcal{C} \equiv D_i E^i - 4\pi\rho. \tag{4.4}$$

Evidently we have $\mathcal{C} = 0$ when the constraint is satisfied. In exercise 2.1 you showed that the evolution equations (4.1) and (4.2), together with charge conservation $\partial_t\rho = -D_i j^i$, also imply that

$$\partial_t\mathcal{C} = 0. \tag{4.5}$$

We say that the constraints are preserved by the evolution equations: if they are satisfied at some initial time, they will be satisfied at all times. The downside is this: in a simulation, numerical error when solving either the constraint or evolution equations may lead to a non-zero constraint violation $\mathcal{C}$; according to (4.5), this constraint violation would then remain even analytically.

When we compared different sets of evolution equations in Section 2.4 we pointed to second-derivative terms other than the Laplace operator as – at least potentially – destroying the desirable properties of a wave equation. In the above equations, the Laplace operator is the first term on the right-hand side of (4.2), $D_j D^j A_i = D^2 A_i$, while the second term, $D_i D^j A_j$, the gradient of the divergence of $\mathbf{A}$, is cause for concern. In order to eliminate this term, consider the following reformulation of Maxwell's equations.[1] We first introduce a new variable

$$\Gamma \equiv D_j A^j; \tag{4.6}$$

this is step (i), as described above, in which a new variable absorbs derivatives of an existing variable.

A brief aside before we proceed. The observant reader will recognize that we can always choose Γ to vanish, namely by imposing the Coulomb gauge $D_i A^i = 0$ and thereby completely eliminating the "term of concern". In general relativity, we can similarly use our coordinate freedom to eliminate the equivalent terms. However, we would rather not use up this freedom at this point since we still need to find coordinates that are suitable for the modeling of black holes and other relativistic objects (more on that in Section 4.2 below). By the same token, we will not impose the Coulomb gauge here but instead will consider Γ as a new independent variable.

---

[1] See Knapp et al. (2002).

Next, therefore, we derive an evolution equation for Γ by taking a derivative of its definition (4.6),

$$\partial_t \Gamma = \partial_t D^i A_i = D^i \partial_t A_i = -D^i E_i - D^i D_i \phi. \tag{4.7}$$

Since the constraint violations $\mathcal{C}$ defined in (4.4) vanish as long as the constraint (4.3) holds, we can now add an arbitrary multiple of $\mathcal{C}$ to the right-hand side of (4.7); this follows step (ii) as described above. Suppose that we add $\eta \mathcal{C}$, where η is a yet-to-be-determined constant. We then have

$$\partial_t \Gamma = -D^i E_i - D^i D_i \phi + \eta \mathcal{C} = (\eta - 1) D_i E^i - D^i D_i \phi - 4\pi \eta \rho. \tag{4.8}$$

Equation (4.8) and the pair

$$\partial_t A_i = -E_i - D_i \phi, \tag{4.9}$$

$$\partial_t E_i = -D^2 A_i + D_i \Gamma - 4\pi j_i \tag{4.10}$$

now form the new system of evolution equations. Equation (4.6) plays the role of a new constraint equation, in addition to the old constraint (4.3).

Exercise 4.1 Show that the constraint violations $\mathcal{C}$ satisfy the equation

$$(-\partial_t^2 + \eta D^i D_i) \mathcal{C} = 0 \tag{4.11}$$

when the fields A_i, E_i, and Γ are evolved with the system (4.8), (4.9), and (4.10), subject to charge conservation $\partial_t \rho = -D_i j^i$.

Exercise 4.1 demonstrates that the constraint violations $\mathcal{C}$ no longer satisfy equation (4.5); instead, they now satisfy equation (4.11). The character of this latter equation depends on the sign of η. For positive η, we see that $\mathcal{C}$ now satisfies a wave equation with the wave speed given by $\eta^{1/2}$. For negative η, however, we cannot expect $\mathcal{C}$ to be well behaved.[2] Evidently, the sign of η is crucially important. Perhaps the most natural choice for η is unity, $\eta = 1$, in which case the divergence of the electric field E^i disappears from equation (4.8) and the constraint

---

[2] One way to see this is as follows. If our problem were subject to periodic boundary conditions, we could consider a Fourier transform of the solution $\mathcal{C} = \mathcal{C}(t, x^i)$, say $\mathcal{C}(t, x^i) = (2\pi)^{-3/2} \int d^3k \exp(-ik_i x^i) \bar{\mathcal{C}}_{\mathbf{k}}(t)$. Inserting this into (4.11) yields an equation for the Fourier coefficients $\bar{\mathcal{C}}_{\mathbf{k}}(t)$, namely $\partial_t^2 \bar{\mathcal{C}}_{\mathbf{k}} + \eta k_i k^i \bar{\mathcal{C}}_{\mathbf{k}} = 0$. If $\eta > 0$, this is a harmonic oscillator equation, and we obtain periodic solutions for the coefficients $\bar{\mathcal{C}}_{\mathbf{k}}$. If, however, $\eta < 0$, the coefficients $\bar{\mathcal{C}}_{\mathbf{k}}$ can grow without bound.

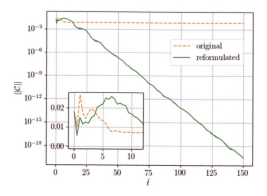

Figure 4.1 The L2-norm $||\mathcal{C}|| \equiv (\int \mathcal{C}^2 d^3x)^{1/2}$ measuring constraint violations in the evolution of an electromagnetic wave, as obtained with both the original (equations 4.1 and 4.2) and the reformulated (equations 4.8–4.10) versions of Maxwell's equations. The results were obtained with the script maxwell.py given in Section B.3.2, with $n_{grid} = 34$ grid points and the outer boundary at $\bar{x}_{out} = 4$ in dimensionless units; see Section B.3.2 for the details and initial data. As one might expect from the discussion in the text, the constraint violations remain constant at late times for the original version and decay exponentially for the reformulated version. (See also Knapp et al. (2002).)

violations propagate at the speed of light. Rather than "sticking around", as in the original Maxwell equations, the constraint violations now propagate like waves and, assuming suitable boundary conditions, can leave the numerical grid. Assuming that a certain fraction of the constraint can leave the grid in each light-crossing time, one might expect an exponential decay. This is exactly what the results shown in Fig. 4.1 demonstrate. Evidently, this is a very desirable property.

In fact, the reader can explore this behavior with the Python script maxwell.py in Section B.3.2. This script serves as an example of an evolution code and allows the reader to experiment with some numerical techniques that are often employed in initial-value problems. The results shown in Fig. 4.1 were produced with this script.

Exercise 4.2   (a) Run the Python script maxwell.py described in Appendix B.3.2 and available at www.cambridge.org/NRStartingfromScratch to evolve the electromagnetic wave initial data (B.55). Plot the data using maxwell_animate.py, also available at www.cambridge.org/NRStartingfromScratch, to reproduce the snapshots shown in Fig. B.1.

(b) Perform simulations with both the original (equations 4.1–4.2) and the reformulated (equations 4.8–4.10) versions of Maxwell's equations. The script maxwell.py produces data files with constraint violations as a function of time. Compare the behavior of constraint violations in the two formulations at late times to confirm the results shown in Fig. 4.1.

(c) Read Appendices B.1 and B.3 for an overview of the numerical techniques employed in constructing `maxwell.py`.

(d) Identify in `maxwell.py` the lines of code that implement the equations in Appendices B.1 and B.3 and that are used to evolve electromagnetic waves.

We encourage readers to work through exercises B.7–B.10 in Appendices B.3.2 and B.4 to explore further some numerical techniques involved in solving evolution equations.

4.1.2 Einstein's Equations

In general relativity we can follow steps very similar to those above to derive a reformulation of the ADM evolution equations

$$\partial_t \gamma_{ij} = -2\alpha K_{ij} + \mathcal{L}_\beta \gamma_{ij}, \tag{4.12}$$

$$\partial_t K_{ij} = \alpha(R_{ij} - 2K_{ik}K^k_{\ j} + K K_{ij}) - D_i D_j \alpha + 4\pi \alpha M_{ij} + \mathcal{L}_\beta K_{ij}, \tag{4.13}$$

hoping to improve their numerical behavior. Various different formulations are currently being used in numerical relativity, and some do not even start with the ADM equations or a 3+1 decomposition at all but instead work with the four-dimensional version of Einstein's equations directly. All formulations share a key ingredient, however, which we can motivate by invoking an analogy with our discussion in Section 4.1.1.

The key point is as follows. Recall that the Ricci tensor R_{ij} harbors second derivatives of the spatial metric. Moreover, we can write the Ricci tensor R_{ij} in different forms; see equations (D.16)–(D.18) of Appendix D. The version that shows the most similarity to the Maxwell equation (4.2) is the second version, equation (2.116),

$$R_{ij} = -\frac{1}{2}\gamma^{kl}\left(\partial_k\partial_l\gamma_{ij} + \partial_i\partial_j\gamma_{kl} - \partial_i\partial_l\gamma_{kj} - \partial_k\partial_j\gamma_{il}\right)$$
$$+ \gamma^{kl}\left(\Gamma^m_{il}\Gamma_{mkj} - \Gamma^m_{ij}\Gamma_{mkl}\right). \tag{4.14}$$

The most important terms for our purposes are the four second-derivative terms. The first of these, $\gamma^{kl}\partial_k\partial_l\gamma_{ij}$, acts like a Laplace operator on the dynamical variable γ_{ij}, as in the first term on the right-hand side of (4.2). Recalling our discussion at the end of Section 2.4, this term could be part of a wave equation for γ_{ij} but that property is destroyed by the remaining three "terms of concern", which involve other permutations of second derivatives that are similar to the gradient of the divergence in the Maxwell equation (4.2).

In (4.2) we rewrote the gradient of the divergence by introducing the new variable Γ in (4.6). In complete analogy, the three "terms of concern" in (4.14) can be rewritten with the help of a new object, namely the trace of the Christoffel symbols,

$$\Gamma^i \equiv \gamma^{jk} \Gamma^i_{jk}. \tag{4.15}$$

In terms of this, the Ricci tensor R_{ij} takes the form

$$R_{ij} = -\frac{1}{2} \gamma^{kl} \partial_k \partial_l \gamma_{ij} + \gamma_{k(i} \partial_{j)} \Gamma^k + \Gamma^k \Gamma_{(ij)k}$$
$$+ 2 \gamma^{lm} \Gamma^k_{m(i} \Gamma_{j)kl} + \gamma^{kl} \Gamma^m_{il} \Gamma_{mkj}, \tag{4.16}$$

the three-dimensional version of (D.18). The first term on the right-hand side is still the Laplace-operator-type second derivative of γ_{ij}, but all other second derivatives of γ_{ij} are now absorbed in first derivatives of Γ^a – in complete analogy with the term $D_i \Gamma$ in (4.10).

We can push the analogy with electrodynamics further by invoking the identity (A.43), proven in exercise A.8, which relates the trace of the Christoffel symbols to the divergence of the metric. Inserting the three-dimensional version of (A.43) into the definition (4.15) we obtain

$$\Gamma^i = -\frac{1}{\gamma^{1/2}} \partial_j \left(\gamma^{1/2} \gamma^{ij} \right), \tag{4.17}$$

where γ is the determinant of γ_{ij}. We now see that Γ^i is directly related to the divergence of the "blue" variable γ_{ij} – in complete analogy to the definition of Γ in (4.6).

It is time for another brief aside. The properties of Γ^i, or rather of its four-dimensional equivalent $^{(4)}\Gamma^a$, were appreciated very early on.[3] In fact, we can choose the $^{(4)}\Gamma^a$ to vanish, which amounts to choosing so-called *harmonic coordinates*, in complete analogy to choosing $\Gamma = 0$ in the Coulomb gauge of electrodynamics. As before, though, we do not yet want to abandon our coordinate freedom and therefore will not yet specify a choice for Γ^i.

Instead, we will regard the Γ^i as new independent variables, meaning that we need to derive new evolution equations. Retracing our steps in equations (4.7) and (4.8) for electrodynamics gives us guidance for what

---

[3] See De Donder (1921); Lanczos (1922).

we need to do: take a time derivative of the expressions (4.17), permute the time and space derivatives, and use the evolution equation for the γ^{ij}, equation (4.12). This will leave us with expressions involving divergences of the extrinsic curvature K^{ij} that are similar to (4.7). Finally we add the momentum constraint (2.119), which eliminates this divergence; as we saw in Section 4.1.1, this step is crucial.

Unfortunately the above steps are complicated by the appearance of the determinant γ in (4.17). This observation motivates adding one more layer of abstraction, namely a conformal transformation, with which we are already familiar from Chapter 3. In the current context it is customary to write the conformal factor ψ in (3.5) as $\psi = e^{\phi}$, so we now have

$$\gamma_{ij} = e^{4\phi}\bar{\gamma}_{ij}. \tag{4.18}$$

Sometimes we refer to ϕ as the *conformal exponent*. Note that we have six independent components on the left-hand side of (4.18) but seven on the right-hand side. Evidently there is still some freedom in choosing the conformal factor – we will return to this shortly.

The main point is the following. We can retrace our steps in Section 3.2.1 (including those that we skipped over) and rewrite physical objects in terms of conformally related objects plus correction terms. We did this explicitly for the Christoffel symbols in (3.9), showing that the physical Christoffel symbols can be written as conformally related Christoffel symbols, computed from derivatives of the conformally related metric as in (3.8), plus derivatives of the conformal factor, now expressed in terms of ϕ. It turns out that we can similarly write the physical Ricci tensor (4.16) as a sum of two terms,

$$R_{ij} = \bar{R}_{ij} + R^{\phi}_{ij}, \tag{4.19}$$

where $\bar{R}_{ij}$ is computed as in (4.16) except with γ_{ij} and Γ^i_{ij} replaced by their conformally related counterparts $\bar{\gamma}_{ij}$ and $\bar{\Gamma}^i_{ij}$:

$$\bar{R}_{ij} = -\frac{1}{2}\bar{\gamma}^{kl}\partial_k\partial_l\bar{\gamma}_{ij} + \bar{\gamma}_{k(i}\partial_{j)}\bar{\Gamma}^k + \bar{\Gamma}^k\bar{\Gamma}_{(ij)k}$$
$$+ 2\bar{\gamma}^{lm}\bar{\Gamma}^k_{m(i}\bar{\Gamma}_{j)kl} + \bar{\gamma}^{kl}\bar{\Gamma}^m_{il}\bar{\Gamma}_{mkj}. \tag{4.20}$$

In (4.19) R^{ϕ}_{ij}, which contains up to second derivatives of ϕ, is given by

$$R^{\phi}_{ij} = -2\left(\bar{D}_i\bar{D}_j\phi + \bar{\gamma}_{ij}\bar{\gamma}^{lm}\bar{D}_l\bar{D}_m\phi\right)$$
$$+ 4\left((\bar{D}_i\phi)(\bar{D}_j\phi) - \bar{\gamma}_{ij}\bar{\gamma}^{lm}(\bar{D}_l\phi)(\bar{D}_m\phi)\right). \tag{4.21}$$

Equation (3.10) in Chapter 3 is the trace of equation (4.19). Note that (4.20) now involves

$$\bar{\Gamma}^i = \bar{\gamma}^{jk}\bar{\Gamma}^i_{jk} = -\frac{1}{\bar{\gamma}^{1/2}}\partial_j\left(\bar{\gamma}^{1/2}\bar{\gamma}^{ij}\right), \tag{4.22}$$

which we refer to as *conformal connection functions*. Here is the beauty: we can now use the freedom in the specification of the conformal factor to choose $\bar{\gamma} = 1$, thereby reducing the conformal connection functions to

$$\bar{\Gamma}^i = -\partial_j\bar{\gamma}^{ij}, \tag{4.23}$$

thereby greatly simplifying the following derivation. We note, however, that choosing $\bar{\gamma} = 1$ makes sense only in Cartesian coordinates, which we will assume in the following.[4]

Taking the determinant of (4.18) we now have

$$\phi = \frac{1}{12}\ln\gamma, \tag{4.24}$$

where γ is the determinant of γ_{ij}. As in Chapter 3 we split the extrinsic curvature into its trace and traceless part,

$$K_{ij} = A_{ij} + \frac{1}{3}\gamma_{ij}K, \tag{4.25}$$

and adopt a conformal transformation for A_{ij}. In Chapter 3 we adopted the conformal rescaling (3.24) because it was particularly useful in the context of the constraint equations; now we are interested in the evolution equations, however, and it makes sense to rescale A_{ij} in the same way as the metric,

$$A_{ij} = e^{4\phi}\tilde{A}_{ij}. \tag{4.26}$$

Note that we will use a tilde here instead of a bar in order to distinguish the two different conformal rescalings, (4.26) as against (3.24).

---

[4] In general it is useful to introduce, as auxiliary objects, a so-called *reference metric* $\hat{\gamma}_{ij}$ together with its associated geometrical objects, i.e. Christoffel symbols, the Riemann tensor, etc. Usually one chooses $\hat{\gamma}_{ij}$ to be the flat metric in whatever coordinate system is used – in spherical polar coordinates, for example, the choice would be the spatial part of (1.21). The equations can then be expressed quite naturally in terms of the *differences* between the physical geometric objects and their reference counterparts; see, e.g., Bonazzola et al. (2004); Shibata et al. (2004); Brown (2009); Gourgoulhon (2012); Montero and Cordero-Carrión (2012); Baumgarte et al. (2013); see also footnote 5 for another advantage of the reference-metric approach.

The dynamical variables are now the conformal exponent ϕ, the conformally related metric $\bar{\gamma}_{ij}$, the mean curvature K, the trace-free part of the extrinsic curvature $\tilde{A}_{ij}$, and the conformal connection functions $\bar{\Gamma}^i$. We can derive evolution equations as follows: for ϕ and $\bar{\gamma}_{ij}$ we take the trace and trace-free parts of (4.12) using equation (4.24) (compare part (a) of exercise 2.26); for K and $\tilde{A}_{ij}$ we do the same for (4.13) (see part (c) of exercise 2.26); and for $\bar{\Gamma}^i$ we follow the steps outlined above.[5] The result of this derivation is what is often referred to as the *Baumgarte–Shapiro–Shibata–Nakamura* formulation,[6] or *BSSN* for short. We list the BSSN equations in Box 4.1.

The BSSN formulation is quite commonly used in numerical relativity and publicly available, for example, through the Einstein Toolkit.[7] It has also been adopted in the Illinois general relativistic magnetohydrodynamics (GRMHD) code.[8] However, it is certainly not the only successful formulation. Another appealing formulation, in the context of simulations in three spatial dimensions without symmetry assumptions, is the *Z4* formulation. Originally derived independently,[9] recent variations and implementations build on the BSSN formalism but have the additional advantage that constraint violations are further suppressed.[10]

Yet another commonly used formulation is the *generalized harmonic* formulation,[11] which Frans Pretorius adopted in his very first simulation of merging black holes.[12] Actually, it is not a reformulation of the ADM equations at all but is instead a four-dimensional formulation that is

---

[5] One complication in these derivations is the fact that several of the dynamical variables now behave as *tensor densities* rather than tensors. A determinant, for example, does not transform from one coordinate system to another like a scalar, since it picks up the square of the Jacobian. We refer to the power of the Jacobian in the coordinate transformation as the *weight* of the tensor density; the determinant, for example, is a scalar density of weight +2, while "regular" tensors are tensor densities of weight zero. Because of (4.18) with $\bar{\gamma} = 1$, the conformal factor is a scalar density of weight 1/6, and $\bar{\gamma}_{ij}$ and $\tilde{A}_{ij}$ are tensor densities of rank 2 and weight $-2/3$. The covariant and Lie derivatives of tensor weights introduce extra terms (see, e.g., Appendix A.3 in Baumgarte and Shapiro (2010)), which need to be taken into account when deriving these equations. Alternatively, the equations can be derived with the help of a reference metric, see footnote 4. Equation (4.24) is then replaced by $\phi = \frac{1}{12} \ln(\gamma/\hat{\gamma})$, where $\hat{\gamma}$ is the determinant of the reference metric – e.g. the flat metric in whatever coordinate system is used. The ratio between two determinants also transforms like a scalar, and we do not need to introduce tensor densities.

[6] See Nakamura et al. (1987); Shibata and Nakamura (1995); Baumgarte and Shapiro (1998).

[7] See www.einsteintoolkit.org.

[8] See Etienne et al. (2010). Many of the GRMHD modules of this code have been ported to the Einstein Toolkit; see Etienne et al. (2015, 2017).

[9] See Bona et al. (2003).

[10] See, e.g., Bernuzzi and Hilditch (2010); Alic et al. (2012).

[11] See Friedrich (1985); Garfinkle (2002); Gundlach et al. (2005); Pretorius (2005b).

[12] See Pretorius (2005a).

Box 4.1 The BSSN equations

The evolution equations in the BSSN formulation are

$$\partial_t \phi = -\frac{1}{6}\alpha K + \beta^i \partial_i \phi + \frac{1}{6}\partial_i \beta^i \tag{4.27}$$

for the conformal exponent ϕ,

$$\partial_t K = -\gamma^{ij} D_i D_j \alpha + \alpha \left(\tilde{A}_{ij} \tilde{A}^{ij} + \frac{1}{3}K^2 \right) + 4\pi\alpha(\rho + S) + \beta^i \partial_i K \tag{4.28}$$

for the mean curvature K,

$$\partial_t \bar{\gamma}_{ij} = -2\alpha \tilde{A}_{ij} + \beta^k \partial_k \bar{\gamma}_{ij} + \bar{\gamma}_{ik}\partial_j \beta^k + \bar{\gamma}_{kj}\partial_i \beta^k - \frac{2}{3}\bar{\gamma}_{ij}\partial_k \beta^k \tag{4.29}$$

for the conformally related metric $\bar{\gamma}_{ij}$,

$$\partial_t \tilde{A}_{ij} = e^{4\phi}\left(-(D_i D_j \alpha)^{\mathrm{TF}} + \alpha(R_{ij}^{\mathrm{TF}} - 8\pi S_{ij}^{\mathrm{TF}}) \right) + \alpha(K\tilde{A}_{ij} - 2\tilde{A}_{il}\tilde{A}^l{}_j)$$
$$+ \beta^k \partial_k \tilde{A}_{kj} + \tilde{A}_{ik}\partial_j \beta^k + \tilde{A}_{kj}\partial_i \beta^k - \frac{2}{3}\tilde{A}_{ij}\partial_k \beta^k \tag{4.30}$$

for the conformally rescaled trace-free part of the extrinsic curvature $\tilde{A}_{ij}$, and

$$\partial_t \bar{\Gamma}^i = -2\tilde{A}^{ij}\partial_j \alpha + 2\alpha \left(\bar{\Gamma}^i_{jk}\tilde{A}^{kj} - \frac{2}{3}\bar{\gamma}^{ij}\partial_j K - 8\pi\bar{\gamma}^{ij} j_j + 6\tilde{A}^{ij}\partial_j \phi \right)$$
$$+ \beta^j \partial_j \bar{\Gamma}^i - \bar{\Gamma}^j \partial_j \beta^i + \frac{2}{3}\bar{\Gamma}^i \partial_j \beta^j + \frac{1}{3}\bar{\gamma}^{li}\partial_l \partial_j \beta^j + \bar{\gamma}^{lj}\partial_j \partial_l \beta^i \tag{4.31}$$

for the conformal connection functions $\bar{\Gamma}^i$. In equation (4.30) the superscript TF stands for the trace-free part of a tensor, e.g. $R_{ij}^{\mathrm{TF}} = R_{ij} - \gamma_{ij}R/3$. We have also split the Ricci tensor R_{ij} into $R_{ij} = \bar{R}_{ij} + R_{ij}^\phi$, where $\bar{R}_{ij}$ and R_{ij}^ϕ are given by equations (4.20) and (4.21). The matter sources ρ, S_{ij}, and S are defined by equations (2.118) and (2.124).

---

derived directly from Einstein's equations. It nevertheless invokes the 3+1 decomposition for the specification of initial data; it also builds on arguments similar to those developed above, except that we are now in a four-dimensional context. Specifically, it also adopts the version (D.18) to express the Ricci tensor in Einstein's equations. Instead of evolving the $^{(4)}\Gamma^a$ freely, however, they are chosen as functions of the coordinates,

$$^{(4)}\Gamma^a = H^a, \tag{4.32}$$

where the freely chosen *gauge source functions* $H^a = H^a(x^b)$, rather than a lapse and shift, now impose the gauge. Choosing $H^a = 0$ results in the *harmonic coordinates* that we mentioned before (see also Section 4.2.3 below) but, since this formalism allows for non-zero H^a, it is referred to as "generalized harmonic". A first-order version of the formulation[13] is used in the SpEC code.[14]

4.2 Slicing and Gauge Conditions

We are now *almost* ready to perform dynamical numerical relativity simulations. One more thing: we need to specify our coordinates, i.e. in a 3+1 decomposition we need to make choices for the lapse function α and the shift vector β^i.

4.2.1 Geodesic Slicing

Why not choose

$$\alpha = 1, \qquad \beta^i = 0, \tag{4.33}$$

the easiest possible choice? For reasons that will become clear in a moment, this choice is called *geodesic slicing* – and the same argument will explain why this is usually a terrible choice.

Recall a few arguments from Section 2.2.3. Exercise 2.10 showed that the acceleration a_a of a normal observer is related to the gradient of the lapse function α by

$$a_a = D_a \ln \alpha. \tag{4.34}$$

For geodesic slicing the acceleration of a normal observer evidently vanishes, meaning that normal observers are freely falling and therefore follow geodesics – hence the name. From equation (2.41) we see that the shift vector β^i describes the relative motion between a coordinate observer, whose worldline connects events with the same spatial coordinates, and a normal observer. With $\beta^i = 0$ the two observers are the same, meaning that, in geodesic slicing, coordinate observers are freely falling.

[13] See Lindblom et al. (2006).
[14] See https://www.black-holes.org/SpEC.html.

But this is bad news! If the spatial coordinates are attached to observers who are freely falling, then any gravitating system will make the spatial coordinates converge, and ultimately intersect, leading to coordinate singularities. For a spherical star or a black hole, for example, all spatial coordinates will fall into the center. In fact, exercise 4.3 below shows that a coordinate singularity will inevitably develop within a finite time once the mean curvature K becomes positive.[15]

> **Exercise 4.3** (a) Use equation (4.28) to show that geodesic slicing implies $\partial_t K \geq 0$ (as long as $\rho + S \geq 0$).
>
> (b) Now neglect the term $\tilde{A}_{ij} \tilde{A}^{ij}$ and assume a vacuum in (4.28) to obtain a differential equation for K alone.
>
> (c) Solve the differential equation of part (b), assuming that $K = K_0 > 0$ at some time $t = t_0$, and find the time t_∞ at which K becomes infinite. In light of the assumptions in part (b), t_∞ serves as an upper limit to the time at which a coordinate singularity will develop.

4.2.2 Maximal Slicing

The arguments of exercise 4.3 suggest that we can control the focusing of coordinate observers by controlling the mean curvature K. For example, we could choose *maximal slicing*,

$$K = 0, \tag{4.35}$$

which we have already encountered in Section 3.3.2 in the context of the constraint equations. There we imposed $K = 0$ on an initial slice only; now we impose $K = 0$ at *all* times, meaning that we also have $\partial_t K = 0$, so that equation (4.28) becomes

$$D^2 \alpha = \alpha \tilde{A}_{ij} \tilde{A}^{ij} + 4\pi \alpha (\rho + S). \tag{4.36}$$

Equation (4.36) for the lapse is an elliptic equation, much like the Poisson equation (1.13), and contains no time derivatives.

Maximal slicing has some very appealing properties, including its *singularity-avoidance* properties. Maximal slices of the Schwarzschild spacetime, for example, avoid the spacetime singularity at areal radius $R = 0$.[16] In regions of strong gravitational fields, the maximal-slicing condition makes the lapse drop to values close to zero – the so-called

---

[15] In a cosmological context, geodesic coordinates are quite commonly used and are referred to as Gaussian normal coordinates. In a Robertson–Walker metric for an expanding universe $K = -3H < 0$, so that the assumptions of part (c) in exercise 4.3 do not apply.

[16] See, e.g., Reinhart (1973); Estabrook et al. (1973).

collapse of the lapse – so that slices are "held back" and no longer advance forward in *proper* time in these regions, even as the coordinate time $t \rightarrow \infty$.[17] This has disadvantages, however, since it leads to so-called *grid stretching*. Quantities like metric coefficients may vary over many decades in the vicinity of a black hole, making it difficult to cover such regions with a sufficient number of grid points when treating a problem numerically.[18] Moreover, imposing maximal slicing entails solving the Poisson-type equation (4.36) for the lapse, which is generally quite costly (see Appendix B.2) and also requires boundary conditions that are sometimes awkward to formulate and impose. For many applications it is therefore easier to deal with a time-evolution equation for the lapse – harmonic slicing provides an example.

4.2.3 Harmonic Slicing

We mentioned above that *harmonic coordinates* result from the requirement that $^{(4)}\Gamma^a = g^{bc(4)}\Gamma^a_{bc} = 0$. Here we will consider a variation of this condition, called *harmonic slicing*, in which we set to zero the time component of $^{(4)}\Gamma^a$ only, and simultaneously choose the shift vector so that it vanishes,

$$^{(4)}\Gamma^0 = 0, \qquad \beta^i = 0. \tag{4.37}$$

In order to obtain a condition for the lapse function α we compute

$$^{(4)}\Gamma^0 = g^{bc(4)}\Gamma^0_{bc} = \frac{1}{|g|^{1/2}}\partial_a\left(|g|^{1/2}g^{a0}\right) = \frac{1}{|g|^{1/2}}\partial_t\left(-|g|^{1/2}\alpha^{-2}\right)$$

$$= -\frac{1}{\alpha\gamma^{1/2}}\partial_t\left(\gamma^{1/2}\alpha^{-1}\right) = 0, \tag{4.38}$$

where we have used the identity (A.43) in the second equality, the identification (2.102) with $\beta^i = 0$ in the third equality, and the relation (2.108) in the fourth. We can therefore write the harmonic slicing condition in the algebraic form

$$\alpha = C(x^i)\gamma^{1/2}, \tag{4.39}$$

where $C(x^i)$ is a constant of integration that may depend on the spatial coordinates. Alternatively we can use (2.126) to obtain a time-evolution equation for the lapse:

$$\partial_t \alpha = -\alpha^2 K. \tag{4.40}$$

A little aside: notice that harmonic slices that are static (so that $\partial_t \alpha = 0$) must also be maximally sliced provided that $\alpha > 0$.

We again consider equation (4.28), to gain some insight into the behavior of harmonic slicing. Consider a vacuum spacetime characterized by a small perturbation of Minkowski spacetime – perhaps a weak gravitational wave traveling through empty space. For Minkowski spacetime (in Minkowski coordinates) we have $\alpha = 1$, $\beta^i = 0$, and $K_{ij} = 0$; to linear order in the perturbation we then write $\alpha \simeq 1 + (\delta\alpha)$ and $K \simeq (\delta K)$, so that the leading-order terms in equation (4.28) are

$$\partial_t(\delta K) \simeq -D^2(\delta\alpha). \tag{4.41}$$

We can now take another time derivative of (4.41), insert the linearized version of the harmonic slicing condition (4.40), and obtain a familiar wave equation,

$$(\partial_t^2 - D^2)(\delta K) = 0. \tag{4.42}$$

As we have discussed before, solutions to the wave equation are well behaved (see e.g. footnote 2 in this chapter) and all is good, at least to linear order. For simulations of black holes, however, it turns out that a variation of harmonic slicing is much better.

4.2.4 1+log Slicing and the Gamma-Driver Condition

We will generalize the condition (4.40) in two ways.[19] First, we will decorate the right-hand side with a function $f(\alpha)$, i.e. we will consider

$$\partial_t \alpha = -\alpha^2 f(\alpha) K. \tag{4.43}$$

Evidently, we recover harmonic slicing for $f(\alpha) = 1$ and geodesic slicing for $f(\alpha) = 0$ (assuming $\alpha = 1$ initially). Formally, maximal slicing corresponds to $f(\alpha) \to \infty$. Redoing the perturbative analysis leading up to equation (4.42) we recognize that we will again obtain a wave equation, and can hence expect well-behaved solutions, but only for positive $f(\alpha)$.

---

[19] See Bona et al. (1995), as well as references therein.

We can now speculate on what might be a good choice for the function f. For nearly flat space, when $\alpha \simeq 1$, we may want f to be close to a constant so that, up to this constant, we recover harmonic slicing. Perhaps we would also like to recover the singularity-avoidance properties of maximal slicing in strong-field regions – i.e. for small values of α we require f to be large. One choice for f that accomplishes both, and that has proven remarkably successful, is $f = 2/\alpha$, which results in

$$\partial_t \alpha = -2\alpha K. \qquad (4.44)$$

Exercise 4.4 below shows why this condition is commonly referred to as the *1 + log slicing* condition.

> **Exercise 4.4** Combine (4.44) with (2.126), integrate the result, and choose a suitable constant of integration to obtain
>
> $$\alpha = 1 + \ln \gamma. \qquad (4.45)$$

As a second generalization of (4.37) we will bring the shift vector β^i back into play. Ever since we adopted equation (4.37) we have assumed that the shift vanishes, but now we will relax that assumption. That means that we have to make a choice for the shift – more on that below – but we will generalize (4.43) and (4.44) to include a shift term. A natural way to do that is to include an "advective" term

$$\left(\partial_t - \beta^i \partial_i\right)\alpha = \alpha n^a \partial_a \alpha = -\alpha^2 f(\alpha) K, \qquad (4.46)$$

where, usually, we adopt $f = 2/\alpha$ again. The "advective version" (4.46) is invariant in the sense that on the left-hand side we are taking a derivative along the direction of the normal vector, which is independent of the shift, while in the "non-advective version" (4.43) we are taking a derivative along the worldline of coordinate observers, which does depend on the shift condition.

> **Exercise 4.5** Consider the Schwarzschild spacetime in the coordinates (1.48), and assume that $\alpha = \bar{r}/(\bar{r} + M)$, $\beta^{\bar{r}} = \bar{r}M/(\bar{r} + M)^2$ and $K = M/(\bar{r} + M)^2$ were found in exercises 2.20 and 2.27. Find a function $f(\alpha)$ such that in equilibrium, when $\partial_t \alpha = 0$, the slicing condition (4.46) is consistent with the $\bar{t} = const$ slices of the coordinate system (1.48).[20]

---

[20] If all goes well you should find that, as $\alpha \to 1$ asymptotically, we have $f \to 0$, thus violating the condition $f > 0$ that we discussed above. In general, this choice therefore does not provide a suitable slicing condition. For Schwarzschild spacetimes, however, this condition can be used anyway, and it provides an extremely simple and powerful test for black-hole simulations.

Now, of course, we also need to supply a condition for the shift vector β^i. Numerous options are possible, but many current codes employ some version of the so-called *Gamma-driver* condition,[21] originally designed to minimize changes in the conformal connection functions $\bar{\Gamma}^i$. Versions of this condition that are commonly implemented in Cartesian coordinates include the coupled equations[22]

$$
\begin{aligned}
(\partial_t + \beta^j \partial_j)\beta^i &= \mu B^i, \\
(\partial_t + \beta^j \partial_j)B^i &= (\partial_t + \beta^j \partial_j)\bar{\Gamma}^i - \eta B^i,
\end{aligned}
\tag{4.47}
$$

and the single equation

$$
(\partial_t - \beta^j \partial_j)\beta^i = \mu \bar{\Gamma}^i,
\tag{4.48}
$$

where μ and η are constants. Note that the above conditions are not invariant, so that one may want to replace the partial derivatives with either a covariant derivative or the Lie derivative in yet another variation.[23]

Equations (4.46) and (4.47) (or 4.48) form evolution equations for the lapse α and shift β^i, i.e. they determine their time derivatives only. In a numerical simulation we also have to choose values for these gauge quantities on the initial time slice. The shift is often chosen to vanish initially, $\beta^i = 0$ (together with $B^i = 0$ if the form 4.47 is adopted), while common choices for the initial lapse are either $\alpha = 1$ or the so-called *precollapsed lapse* $\alpha = \psi^{-2}$, where $\psi = \exp(\phi)$ is the initial conformal factor.

The $1 + \log$ slicing condition for the lapse together with a Gamma-driver condition for the shift have been used in many numerical simulations. A common name for this combination is the *moving-puncture gauge*. This name will make more sense in the context of black-hole simulations, where this gauge condition reveals its true magic – and therefore we will postpone a discussion until the next chapter.

---

[21] See Alcubierre and Brügmann (2001).
[22] See, e.g., Alcubierre et al. (2003); van Meter et al. (2006); Thierfelder et al. (2011).
[23] See Brown (2009).

5

Numerical Simulations of Black-Hole Binaries

In the previous chapters we assembled the pieces needed to perform a numerical relativity calculation in a vacuum spacetime. In particular we sketched how to construct initial data and how to evolve these data forward in time. Here we will describe how, when these pieces are pulled together, they can be employed for an important application, namely the simulation of the inspiral and merger of binary black holes, one of the most promising sources of gravitational waves. We will describe different approaches that have been adopted in these calculations, review some unique astrophysical effects that have been discovered, and discuss how these simulations have been used to aid in the detection and interpretation of gravitational wave signals.

5.1 Binary Black Holes and Gravitational Waves

Numerical relativity has been applied to many problems and has celebrated many successes. It would be beyond the scope of this volume to discuss all of them,[1] but we would be remiss to not mention some milestones in the development of the field. These include the beginnings of numerical relativity with the "geometrodynamics" simulations of Hahn and Lindquist,[2] which even predated the coining of the term "black hole", the first simulations of gravitational collapse by May and White,[3] simulations of the head-on collisions of black holes by Smarr[4] and of relativistic clusters and their collapse and mergers by Shapiro

---

[1] See, e.g., the textbooks by Baumgarte and Shapiro (2010) and Shibata (2016) for more comprehensive discussions; see also Cardoso et al. (2015) for a review of numerical relativity applications to new physics frontiers.
[2] See Hahn and Lindquist (1964).
[3] See May and White (1966).
[4] See Smarr (1979).

and Teukolsky,[5] the discovery of critical phenomena in gravitational collapse by Choptuik,[6] and the first simulations of binary neutron-star mergers by Shibata.[7] Nevertheless, for many years one of the holy grails of numerical relativity was the simulation of the inspiral, coalescence, and merger of binary black holes in quasicircular orbit, and that is what we will focus on here.

Simulating binary-black-hole coalescence was an important goal in and of itself, but it also served an even larger goal of the gravitational physics community: namely the identification and interpretation of gravitational wave signals to be detected by LIGO and other gravitational wave observatories. By now both goals have become spectacular reality. Perhaps no figure documents this achievement more beautifully than Fig. 1 of Abbott et al. (LIGO Scientific Collaboration and Virgo Collaboration) (2016a), the paper that announced the first ever direct detection of gravitational waves, an event called GW150914. The figure,

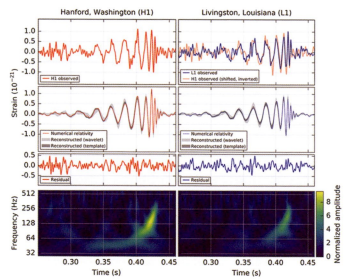

Figure 5.1 Figure 1 of Abbott et al. (LIGO Scientific Collaboration and Virgo Collaboration) (2016a), announcing the first ever direct detection of gravitational radiation. The figure shows the observed gravitational wave signal, called GW150914, together with a signal computed from a numerical relativity simulation of two merging black holes with masses of 36 and 29 solar masses. The top panels show the strain as a function of time, where the strain is directly related to the perturbations h_{ab} that we introduced in Section 1.3.2. The bottom panels show time–frequency representations of the data, which are similar to those of Fig. 1.9.

[5] See, e.g., Shapiro and Teukolsky (1985a,b, 1992).
[6] See Choptuik (1993); see also Gundlach and Martín-García (2007) for a comprehensive review of critical phenomena in gravitational collapse.
[7] See Shibata (1999a,b).

reproduced as Fig. 5.1, shows the gravitational wave signals detected on September 14, 2015 by the LIGO detectors in Hanford, Washington, and Livingston, Louisiana. This stunning discovery was made almost exactly one hundred years after Einstein first predicted the existence of gravitational waves. A comparison of the observed signal with numerical relativity predictions establishes the signal as having been emitted by two black holes with masses of approximately 36 and 29 solar masses that merged about a billion years ago at a distance of about 410 Mpc. The total energy emitted in the form of gravitational waves was a whopping 3 $M_\odot c^2$ (approximately 2×10^{54} ergs), at a rate larger than the combined power of all light radiated by all stars in the observable universe. This very first measured signal was so strong that it clearly emerges from the noise of the detector – truly a gift from nature!

We anticipated in Section 1.3.2 that, in contrast with binary neutron star mergers (see Fig. 1.9), the transition from inspiral to merger should occur within the detectors' most sensitive range for black-hole binaries in the mass range of GW150914. Quite remarkably, this transition is easily visible in Fig. 5.1. The signal starts in the inspiral phase, during which the emission of gravitational waves leads to a loss of energy and angular momentum, hence a shrinking of the binary separation, and thus an increase in both the orbital and gravitational wave frequency (as discussed in Section 1.3.2). The gravitational wave signal peaks at a frequency similar to our crude estimate (1.77) just as the black holes merge.

Exercise 5.1 (a) Estimate the peak frequency f_{GW}^{crit} from Fig. 5.1 and then use (1.77) to obtain a crude estimate of the binary's chirp mass $\mathcal{M}$. *Hint:* Your estimate is likely to be somewhat smaller than the result $\mathcal{M} \simeq 30 M_\odot$ reported in Abbott et al. (LIGO Scientific Collaboration and Virgo Collaboration) (2016b).

(b) Now estimate the peak amplitude h^{crit} from Fig. 5.1 and substitute this and your result from part (a) into (1.78) to obtain a crude estimate for the distance r to the binary. *Hint:* The analysis of Abbott et al. (LIGO Scientific Collaboration and Virgo Collaboration) (2016b), which also takes into account cosmological effects, results in a value of 410^{+160}_{-180} Mpc, somewhat larger than your estimate is likely to suggest.

The merger forms a single black-hole remnant that is not immediately in stationary equilibrium but, rather, is distorted. Distorted black holes oscillate, but since these *quasi-normal oscillations* again emit gravitational radiation, they are exponentially damped – this process therefore allows distorted black holes to settle down into stationary equilibrium, i.e. a rotating Kerr black hole (see Fig. 1.10 for a sketch). This quasi-normal *ringdown*, which has unique normal-mode frequencies and damping coefficients for a given mass and spin, is also clearly visible in the gravitational wave signal shown in Fig. 5.1.

Weaker signals[8] are hidden in the noise, however, and cannot be seen with the naked eye (or, in audio-renderings, heard with the naked ear). In this case we have to rely on sophisticated matched-filtering data analysis, which requires accurate templates of theoretical gravitational wave signals and again requires numerical relativity simulations.[9] Clearly, solving the binary-black-hole problem is of crucial importance from the perspective of gravitational wave detection.

While the goal of simulating merging black holes was important, it was also elusive. In Chapter 4 we already discussed issues with the numerical stability of different formulations of the equations, as well as the ambiguities associated with coordinate freedom. As an additional complication, the evolving of black holes also requires techniques for dealing with black-hole singularities. Solving this problem turned out to be harder than expected when dealing with binary mergers in quasi-circular orbit and took a longer time than anticipated, resisting both the individual and concerted efforts of numerous groups for many years. Perhaps this made the solution, once it was finally found, even more remarkable.

The first numerical simulation of the inspiral, merger, and ringdown of two black holes was presented by Frans Pretorius[10] at a conference in Banff in April 2005. We show some snapshots from his simulations in Fig. 5.2. At early times, the binary separation shrinks as the two black holes orbit each other and the black-hole horizons, the interiors of which are blacked out in the figure, become increasingly distorted. At some point the two horizons merge, resulting in a single distorted black hole. This distorted black hole then performs the damped quasi-normal ringing described above and ultimately settles down into a Kerr black hole. Not too long after Pretorius' remarkable announcement, two groups, one led by Manuela Campanelli and Carlos Lousto at

---

[8] This includes binary-black-hole signals that have already been detected (see Abbott et al. (LIGO Scientific Collaboration and Virgo Collaboration) (2018) for a catalogue of signals detected in the O1 and O2 observing runs, as well as https://www.ligo.caltech.edu for updated lists, including O3, etc.) and also anticipated future events.

[9] Gravitational wave template catalogues are typically built using analytical phenomenological expressions that incorporate post-Newtonian expansions and are calibrated using numerical relativity simulations; see Section 5.4 below.

[10] See Pretorius (2005a).

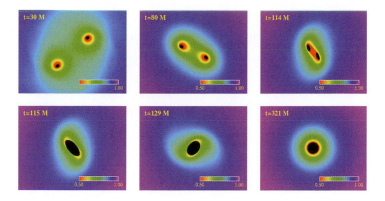

Figure 5.2 Snapshots of the first successful simulation of a binary-black-hole inspiral and merger by Pretorius (2005a). The color coding represents the lapse function α, and the black areas represent the black-hole interiors.

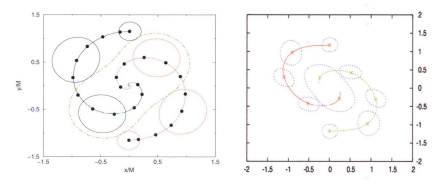

Figure 5.3 Trajectories of the punctures and horizons in the black-hole simulations of Campanelli et al. (2006a) (left-hand panel) and Baker et al. (2006) (right-hand panel). In the left-hand panel, the points represent the location of the punctures at times $t = 0, 2.5M, 5M$, etc., where M is the total mass of the spacetime. Horizons are shown at times $t = 0, t = 10M$ and $t = 18.8M$. The common horizon, marked by the dashed green line, forms at $t = 18.8M$, just after the punctures complete a half orbit. In the right-hand panel, the locations of the apparent horizons are shown at times $t = 0, 5M, 10M, 15M$, and $20M$.

Brownsville, Texas,[11] and the other, led by John Baker and Joan Centrella at NASA Goddard Space Flight Center,[12] presented similarly spectacular simulations of black-hole mergers at a conference at Goddard in the fall of 2005 (see Fig. 5.3). Quite amazingly, both groups

[11] See Campanelli et al. (2006a).
[12] See Baker et al. (2006).

had unknowingly pursued approaches that were independent of that of Pretorius, but very similar to each other, and had succeeded almost simultaneously.

5.2 Black-Hole Excision and the Moving-Puncture Method

In his simulations, Pretorius adopted the generalized harmonic formalism that we described at the end of Section 4.1.2, and he used *black-hole excision* to handle the singularities at the center of the black holes. The key idea of black-hole excision[13] is that no physical information can propagate from the inside of a black hole to the exterior – so why bother including the interior of a black hole in a numerical simulation in the first place? The idea can be implemented by excising, i.e. completely ignoring, at least some grid points inside a black-hole horizon. While this sounds quite simple conceptually, it also brings with it some subtleties – for example, when black holes move through a computational mesh, grid points that re-emerge from a black hole have to be assigned valid function values. The Brownsville and Goddard groups, however, used what is now commonly referred to as the *moving-puncture* method, since it extends earlier attempts to generalize the puncture method for black-hole initial data (see Section 3.4) to moving black holes.[14] Specifically, both groups adopted a version of the BSSN formulation given in Box 4.1 together with what has now become the "standard gauge", or moving-puncture gauge, namely the $1+\log$ slicing condition (4.46) for the lapse and a Gamma-driver condition (4.47) or (4.48) for the shift. Moreover, both groups found that, with this combination, there was no need for black-hole excision or any other way of treating the black-hole singularities – the singularities appeared to take care of themselves!

Much insight into this peculiar behavior, which seemed almost too good to be true, came from applying the moving-puncture method to the Schwarzschild spacetime.[15] These studies showed that, using $1+\log$ slicing, the evolution will settle down to a slicing of the Schwarzschild spacetime in which the spatial slices feature a *trumpet* geometry. Trumpet geometries are named for their appearance in an embedding

---

[13] The original idea of black-hole excision is attributed to William Unruh, as quoted in Thornburg (1987); see Seidel and Suen (1992); Scheel et al. (1995) for some early implementations.
[14] See also Brügmann et al. (2004).
[15] See Hannam et al. (2007b); Brown (2008); Hannam et al. (2008).

diagram (we have already encountered an example in Fig. 1.7) and they are characterized by two key properties[16] in the vicinity of the origin $\bar{r} = 0$, referred to as the *puncture* (which, in Fig. 1.7, would be "all the way down").

First, the proper length of a circle of (coordinate) radius $\bar{r}$ goes to a non-zero, finite value as $\bar{r} \to 0$. In the embedding diagram of Fig. 1.7 this is represented by the fact that, as $\bar{r} \to 0$, the diagram asymptotes to a cylinder of finite radius – the proper length of a circle with radius $\bar{r} \to 0$ corresponds to the circumference of this cylinder. This means that the spatial slice ends at a finite value of the areal (or circumferential) radius R; in particular, therefore, the curvature singularity at $R = 0$ is not part of these slices: the slicing condition has "excised" the black-hole singularity for us, without any need to do it "by hand".

Second, the puncture at $\bar{r} = 0$ is at an infinite proper distance from any point with $\bar{r} > 0$. In the embedding diagram of Fig. 1.7 this is represented by the fact that the point $\bar{r} = 0$ is "all the way down", i.e. at an infinite distance from the black-hole horizon.

Trumpet slices are not unique and can be reproduced using different versions of the 1+ log slicing condition. The "non-advective" 1+ log condition (4.44), for example, will result in maximally sliced trumpet slices at late times.[17] Similarly, exercise 4.5 shows that the analytical Schwarzschild trumpet slices of exercise 1.6 are consistent with (4.46) for a specific choice of $f(\alpha)$. The latter slices provide a particularly simple analytical framework for exploring the two properties discussed above. As it turns out, both are a consequence of the fact that the conformal factor ψ behaves as $\psi \simeq \bar{r}^{-1/2}$ as $\bar{r} \to 0$.

Exercise 5.2 (a) Revisit exercise 1.6 and compute the proper length $\mathcal{L}$ of a circle at coordinate label $\bar{r}$ in the equatorial plane around the origin of the coordinate system (1.48).[18] Note that $\mathcal{L}$ remains greater than zero as $\bar{r} \to 0$. This implies that the coordinates of exercise 1.6 do not reach the spacetime singularity at $R = 0$.

(b) Compute the proper distance $\mathcal{D}$ between the origin at $\bar{r} = 0$ and a point at $\bar{r} = \epsilon$. Show that $\mathcal{D}$ is infinite for any $\epsilon > 0$.

---

[16] See Dennison et al. (2014) for a characterization of trumpet geometries in the absence of spherical symmetry.

[17] Note that $K \to 0$ as $\partial_t \alpha \to 0$ with $\alpha > 0$; see Hannam et al. (2007a); Baumgarte and Naculich (2007).

[18] Either carry out the integration or simply recognize that, by the definition of R,
$\mathcal{L} = 2\pi R = 2\pi(\bar{r} + M)$.

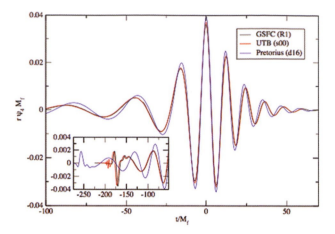

Figure 5.4 A comparison of gravitational waveforms from the three different simulations of Baker et al. (2006) (labeled as GSFC), Campanelli et al. (2006a) (UTB) and Pretorius. Unlike in Pretorius (2005a), Pretorius adopted the initial data of Cook and Pfeiffer (2004) for these simulations (see also Buonanno et al. (2007)). The wave amplitude is measured as a Weyl curvature scalar ψ_4, which can be computed from the second time derivatives of the perturbations h_{ab} introduced in equation (1.55) of Section 1.3.2. (Figure from Baker et al. (2007).)

Both properties are crucial for the well-being of numerical simulations. The first property ensures that no *spacetime* singularity is present on the evolved slices. The puncture still features a *coordinate* singularity (for instance, an $\bar{r}^{-1/2}$ divergence in the conformal factor) but, since this point is at an infinite proper distance from all other points, its effects are very limited.[19] In practice, the singularity at $\bar{r} = 0$ will affect a few grid points in its neighborhood but as long as the resolution is sufficiently high, and all these grid points are inside a black-hole horizon, the solution will converge as expected in the exterior of the black hole. Quite remarkably, the moving-puncture method does not require that anything special be done to accommodate black holes!

Gravitational waveforms predicted from these first simulations of binary-black-hole mergers are shown in Fig. 5.4. Quite reassuringly, this direct comparison shows that the results of Campanelli et al. (2006a), Baker et al. (2006), and Pretorius (2005a) agree quite well.

---

[19] As long as this point does not happen to lie on an actual grid point. Alternatively, some groups evolve the inverse of some power of the conformal factor rather than the conformal factor itself or the conformal exponent. Note also that the shift β^i at the puncture is minus the puncture velocity in a moving-puncture gauge.

The small differences result from the use of differing initial data; the first two groups adopted the puncture method described in Section 3.4, while Pretorius used the conformal thin-sandwich data of Cook and Pfeiffer (2004), which we briefly discussed in Section 3.5 (see also Buonanno et al. (2007)). While all three groups adopted equal-mass binaries, the different approaches in the construction of initial data lead to black holes that carry slightly different spins, binaries with different residual eccentricities, and spaces with slightly different gravitational wave content,[20] which then result in slightly different waveforms. Nevertheless, the waveforms shown in Fig. 5.4 clearly show the general features that we have discussed before and have seen already in Fig. 5.1: an increase in both frequency and amplitude during the inspiral, and a transition to the exponentially decaying quasi-normal ringing during the merger.

Clearly, these first successful simulations of binary-black-hole mergers marked a true breakthrough in the field of numerical relativity. As well as being remarkable calculations in and of themselves, they also seemed to open flood gates, allowing different groups to explore the full parameter space of binary-black-hole mergers. As it turns out, this led to some surprising and unexpected results, with important astrophysical consequences.

5.3 Orbital Hang-up, Black-Hole Recoil, and Spin Flips

The first simulations of binary black holes all considered equal-mass binaries with non-spinning black holes in quasi-circular orbits. In general, however, the companions in black-hole binaries may have different masses and spins and may even inspiral and merge from non-circular initial orbits. These different possibilities can lead to a host of interesting effects.

Consider, for example, a quasi-circular binary in which both black holes carry a spin that is aligned with the orbital angular momentum. The angular momentum of the merger remnant is then the sum of both spins plus the orbital angular momentum, minus whatever angular momentum has been carried away by the emitted gravitational radiation.

---

[20] Including unwanted residual "junk radiation" that is typically present when solving the initial value equations for binaries at finite separation.

Once the remnant has settled down to a Kerr black hole, its angular momentum must satisfy the Kerr limit

$$J \leq M^2, \tag{5.1}$$

where M is the remnant's mass. For their merger remnant to satisfy this constraint, binaries with large spins that are aligned with the orbital angular momentum have to emit more angular momentum than binaries with smaller spins or with spins that are anti-aligned with the orbital angular momentum. Accordingly, the inspiral of the former proceeds more slowly, giving the binary more time to radiate away angular momentum, than that of the latter. This effect, caused by spin–orbit coupling, is called *orbital hang-up*.[21]

Another effect results from the breaking of the symmetry of the binary. Each black hole in a binary orbit is constantly accelerating, and, similarly to an accelerating charge, emits linear momentum. In a symmetric binary the emission of linear momentum from one black hole is canceled exactly by that from its companion, but for an asymmetric binary there is a net loss of linear momentum in outgoing gravitational waves that accumulates during the course of the inspiral. This leads to a *black-hole recoil*, or *black-hole kick*, to conserve total linear momentum. For non-spinning binaries, for which a mass ratio different from unity breaks the symmetry, the recoil can reach speeds of up to about 175 km/s.[22] For binaries with spinning black holes, however, the recoil can reach surprisingly high speeds of up to about 5000 km/s.[23] Since this speed exceeds the escape speed from even the largest galaxies, this has raised the possibility that black-hole recoil might be observable, in the form of supermassive black holes moving away from the centers of their recently merged host galaxies.[24] The exact recoil speed, however, depends sensitively on the orientation of the black-hole spins. The largest kicks are found for anti-aligned spins that are almost perpendicular to the orbital angular momentum;[25] this configuration, even for a random distribution of spins, is not all that likely. Since interactions with accreting gas are expected to align the spins with

---

[21] See, e.g., Campanelli et al. (2006b); Healy and Lousto (2018).
[22] See Herrmann et al. (2007); González et al. (2007).
[23] See, e.g., Campanelli et al. (2007a); González et al. (2007); Campanelli et al. (2007b); Lousto and Zlochower (2011); see also Zlochower and Lousto (2015) and references therein for fitting formulae for the remnant mass, spin, and recoil speed.
[24] See Komossa (2012) for a review; see also Lousto et al. (2017); Chiaberge et al. (2018) for discussions of the more recent recoil candidate QSO 3C 186.
[25] Lousto and Zlochower (2011).

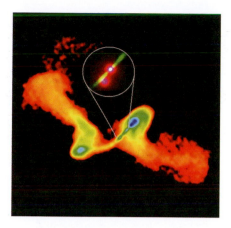

Figure 5.5 An X-shaped radio jet in the radio galaxy NGC 326. The outer part of the image shows jets that point in the upper left and bottom right directions, while in the inner parts, corresponding to more recent emission, the jets point into the bottom left and upper right directions. This change of direction can be explained by a flip in the black-hole spin that reorients the direction of the jets. (Image from National Radio Astronomy Observatory/AUI, observers Matteo Murgia et al., STScI.)

the orbital angular momentum, the likelihood that a perpendicular configuration is realized in nature, with a merger then resulting in a large recoil, depends strongly on the astrophysical environment and history.[26]

Another phenomenon with interesting astrophysical consequences is *spin flip*. A black hole in a binary may have a spin pointing in a certain direction. As we said before, the merger remnant's angular momentum is the sum of the orbital angular momentum and the two black holes' spins, minus whatever angular momentum was carried away by gravitational waves. Therefore, the remnant's spin may point in a direction that is completely different from either of the progenitors' spins.[27] This, too, may have observational consequences, since jets of radiating gas and magnetic fields are expected to be emitted along the black hole's spin axis. Figure 5.5 shows an example of a so-called "X-shaped" radio jet, in which the direction of the jet emission appears

---

[26] See, e.g., Bogdanovic et al. (2007); see also Blecha et al. (2016) for a more recent discussion in the context of cosmological Illustris simulations.
[27] Instabilities in the spin orientation may also lead to spin flips during the inspiral; see Gerosa et al. (2015); Lousto and Healy (2016).

to have changed suddenly at some point in the past. One very plausible explanation for this phenomenon is a binary merger and spin flip.[28]

5.4 Gravitational Wave Catalogs

Finally, we return to one of the original motivations for simulating black-hole mergers, namely the assembly of gravitational wave catalogs to be used in the data analysis of gravitational wave observations. While the advances in numerical relativity and the above simulations represent a crucial step towards constructing such gravitational wave templates, the reality is that numerical relativity simulations alone cannot provide all the data that are needed for an effective data analysis. The reason is that these simulations are quite intensive computationally. On the one hand it is expensive to simulate an individual merger, at sufficient accuracy, through a large number of orbits and hence gravitational wave cycles; on the other hand it would be prohibitively expensive to cover the entire eight-dimensional parameter space of binary-black-hole systems (described by the mass ratio, magnitudes, and orientations of the two black-hole spins, and the orbital eccentricity) with sufficient resolution. While individual groups have assembled gravitational wave catalogs[29] – we show an example in Fig. 5.6 – and while there have been targeted simulations to model specific observed gravitational wave signals,[30] much effort has thereby gone into combining numerical relativity results with analytical results from post-Newtonian theory.

Efforts to use numerical relativity simulations in gravitational wave data analysis predated the first observation of gravitational waves. The "Numerical INJection Analysis" (NINJA) collaboration[31] studied the sensitivity of gravitational wave-search and parameter-estimation algorithms by collecting numerically generated gravitational waveforms from ten different numerical relativity groups and injecting these into simulated data sets designed to mimic the response of gravitational

---

[28] See, e.g., Merritt and Ekers (2002); see also Gopal-Krishna et al. (2012) and references therein for alternative models, such as black-hole precession, to explain X-shaped and winged radio jets.

[29] See https://data.black-holes.org/waveforms/index.html for catalogs of the Caltech/Cornell/CITA collaboration, Mroué et al. (2013), and http://ccrg.rit.edu/~RITCatalog/ for those of the group at Rochester Institute of Technology, Healy et al. (2019).

[30] See, e.g., Healy et al. (2018).

[31] Aylott et al. (2009).

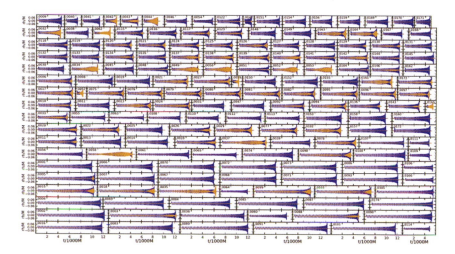

Figure 5.6 A subset of the gravitational waveforms catalogued by the Caltech/Cornell/CITA collaboration, where the two different colors represent the two different polarization state (see Mroué et al. (2013) for details).

wave detectors. Follow-up studies[32] improved this analysis by using real detector data and so-called "hybrid waveforms", which can match numerical relativity results for the late inspiral waveform with post-Newtonian results that accurately describe the early inspiral.

In a related effort, the Numerical-Relativity–Analytical-Relativity (NRAR) collaboration[33] was aimed at using numerical relativity simulations to develop and calibrate accurate analytical gravitational wave templates. The NRAR collaboration considered two families of such analytical wave templates, which later played a crucial role in the first detections of gravitational radiation.

One of these families is based on the so-called Effective-One-Body (EOB) formalism, which describes the binary inspiral by combining strong-field results from the test-particle limit with weak-field, perturbative, results obtained from post-Newtonian approximations.[34] In this approach, the unknown higher-order post-Newtonian terms are calibrated by comparing with numerical relativity (NR) simulations. Gravitational wave templates based on this approach, usually referred

---

[32] Ajith et al. (2012); Aasi et al. (The LIGO, VIRGO and NINJA-2 Collaborations) (2014).
[33] Hinder et al. (2013).
[34] See Buonanno and Damour (1999) and numerous follow-up calculations.

to as EOBNR waveform models, have been extended to include both non-precessing and precessing black-hole spins.[35]

A second family of analytical wave templates is based on a phenomenological description of the inspiral/merger/ringdown phase and is therefore commonly referred to as IMRPhenom. This approach is based on the hybrid waveforms that we discussed above, which combine post-Newtonian descriptions for the early inspiral with numerical relativity simulations of the late inspiral, merger, and ringdown.[36] Later extensions of this approach have also included both non-precessing and precessing spins.[37]

Both families of analytical wave templates have been employed in the analysis of gravitational wave signals – see, e.g., Table I of Abbott et al. (LIGO Scientific Collaboration and Virgo Collaboration) (2016b) for the parameters that characterize the source of GW150914. Together with the use of targeted simulations (one of which is included in Fig. 5.1), this underscores the importance of numerical relativity and of the techniques that we have described in this volume for the emerging field of gravitational wave astronomy.

The detection of the gravitational wave signal GW170817 defines a larger challenge for numerical simulations. This source was the first detection of gravitational waves from a binary neutron star inspiral, as shown in Fig. 1.9.[38] The neutron star masses are estimated to be 1.36–$1.60M_\odot$ for the primary and 1.17–$1.36M_\odot$ for the secondary, assuming that their spins are small. The merger was also identified by the fact that it gave rise to a short gamma-ray burst, GRB 170817A, and a kilonova AT 2017gfo, marking the first joint detection and analysis of gravitational and electromagnetic radiation. In this book we have restricted ourselves to the treatment of vacuum spacetimes in general relativity, i.e. to the treatment of the gravitational fields only, which is sufficient for the modeling of black holes and their gravitational wave emission. However, we have developed this treatment of the gravitational fields by drawing heavily on analogies with scalar and electromagnetic fields, both of which can appear as matter sources in Einstein's equations. In order to give the reader a taste of how the equations of motion for any matter field follow from Bianchi's identities (2.97), and how the matter source terms in the 3+1 decomposition of

---

[35] See, e.g., Taracchini et al. (2014); Pan et al. (2014).
[36] Ajith et al. (2008).
[37] See, e.g., Ajith et al. (2011); Khan et al. (2018).
[38] See Abbott et al. (LIGO Scientific Collaboration and Virgo Collaboration) (2017).

Einstein's equations are computed, we have included in Appendix C a brief sketch of these topics for both scalar and electromagnetic fields. Simulating neutron stars and their mergers, together with their ejecta and afterglows, requires accurate prescriptions for a host of processes and phenomena, including relativistic magnetohydrodynamics, realistic equations of state, and nuclear reaction networks, as well as electro-magnetic radiation and neutrino transport.[39] While these subjects go well beyond the scope of this book, we hope that some readers may be sufficiently curious and inspired to learn about the above, to contribute to the solution of these problems, and thereby to advance an exciting field of research.

---

[39] See, e.g., Shibata et al. (2017); Radice et al. (2018); Rezzolla et al. (2018); Ruiz et al. (2018); Gieg et al. (2019) for early numerical relativity simulations aimed at modeling GW170817.

Epilogue

As we explained in the Preface, we have not aimed for a complete or comprehensive introduction to numerical relativity in this volume, and we are quite confident that we have succeeded in that regard! Instead, we have restricted ourselves to a small set of topics involving the construction of vacuum initial data and their dynamical evolution, confining our discussion to black holes and gravitational waves.

Even for these subjects, however, our treatment is still incomplete. For a start, we have skipped some details and derivations, opting instead for a less technical approach that we hope is more accessible to a broader audience and that perhaps provides a more intuitive introduction to the key quantities and equations. A curious reader can, of course, fill in all the details by consulting other books on numerical relativity.[1] Perhaps more importantly, we have described approaches for solving only the constraint and evolution equations but have not discussed the all-important diagnostic quantities that allow us to calibrate our simulations and to analyze and interpret their results. Such diagnostics include horizon finders, which locate black-hole horizons and provide local measures of the masses and angular momenta of black holes, and gravitational wave extraction techniques, which probe any outgoing gravitational waves emitted by a source. We refer to footnote 1 for texts giving treatments of these and other diagnostics.

By restricting ourselves to vacuum spacetimes, we have obviously omitted a treatment of sources of gravitation arising from any matter and/or nongravitational fields that may be present in the spacetime. An exception is found in Appendix C, where we describe the stress–energy

---

[1] See Bona and Palenzuela (2005); Alcubierre (2008); Bona et al. (2009); Gourgoulhon (2012); Baumgarte and Shapiro (2010); Shibata (2016).

tensors for scalar and electromagnetic fields and use them to derive the equations of motion for these fields. Apart from that discussion, however, we have dealt only with the left-hand side of Einstein's equations, i.e. the "geometric" side encapsulating the gravitational field, and have set the right-hand side, which accounts for matter and non-gravitational fields, to zero, as is appropriate for a vacuum spacetime. But in the presence of such sources we again need to insert their stress–energy tensors on the right-hand side of the equations, as well as to solve the equations of motion for these sources simultaneously with Einstein's field equations. The recent detections of gravitational radiation provide dramatic examples: modeling GW150914, a merging binary black hole system, only requires methods for handling vacuum spacetimes, while modeling GW170817, a merging binary neutron star system, together with its counterpart electromagnetic emission, brings into play a host of additional physics that must be accounted for on the right-hand side of Einstein's equations. Treating this physics typically entails working with the equations of relativistic magnetohydrodynamics, a realistic equation of state for hot dense nuclear matter, nuclear reaction networks, and electromagnetic and neutrino radiation transport. Some of the methods that are needed are already well established for computational implementation, while others still require significant approximations to perform practical simulations, pending algorithmic and/or computer hardware breakthroughs. Again we refer to footnote 1 for textbook treatments of some of these topics, as well as the current literature to track the most recent developments.

Before closing we reiterate that several numerical relativity codes are publicly available, ready to be downloaded, and can be used to evolve black holes, neutron stars, compact binary mergers, cosmological models, and other phenomena involving strong gravitational fields. Examples of such open-source codes include the Einstein Toolkit, available at www.einsteintoolkit.org. As we discussed in the Preface, we hope that this volume may help to make sense of some quantities and equations encountered in those codes. We further hope that, by following this introductory text, readers will be able to perform numerical relativity simulations using these or other tools with greater confidence, awareness, and dexterity. Finally, we hope that inspired readers will explore the subject in more depth and help to advance our rapidly expanding understanding of general relativity and relativistic astrophysics.

Appendix A: A Brief Review of Tensor Properties

A.1 Expansion into Basis Vectors and One-Forms

A tensor is a geometric object and has an intrinsic meaning that is independent of the coordinates or basis vectors chosen to represent it. A vector **A**, for example, has a certain length and points in a certain direction independently of the basis vectors in which we express this vector.

We can expand any tensor either in terms of basis vectors $\mathbf{e}_a$ or basis one-forms $\boldsymbol{\omega}^a$, as we will now discuss.[1] For a tensor of rank n we will need a grand total of n basis vectors or one-forms. A vector is a rank-1 tensor that can be expanded in basis vectors, e.g.

$$\mathbf{A} = A^a \mathbf{e}_a. \tag{A.1}$$

This expression deserves several comments. For a start, we have used the Einstein summation rule, meaning that we sum over repeated indices. The "upstairs" index a on A^a refers to a *contravariant* component of **A**. The index a on the basis vector $\mathbf{e}_a$, however, does *not* refer to a component of the basis vector – instead it denotes the name of the basis vector (e.g. $\mathbf{e}_x$ or $\mathbf{e}_1$ is the basis vector pointing in the x direction). If we wanted to refer to the bth component of the basis vector $\mathbf{e}_a$, say, we would write $(\mathbf{e}_a)^b$.

We now write the dot product between two vectors as

$$\mathbf{A} \cdot \mathbf{B} = (A^a \mathbf{e}_a) \cdot (B^b \mathbf{e}_b) = A^a B^b \, \mathbf{e}_a \cdot \mathbf{e}_b. \tag{A.2}$$

---

[1] Many textbooks on general relativity treat the subjects of this appendix much more thoroughly and rigorously. The treatments in Misner et al. (2017), Hawking and Ellis (1973), Schutz (1980), Wald (1984), and Carroll (2004) are particularly detailed.

Defining the metric as

$$g_{ab} \equiv \mathbf{e}_a \cdot \mathbf{e}_b \tag{A.3}$$

we obtain

$$\mathbf{A} \cdot \mathbf{B} = A^a B^b g_{ab}. \tag{A.4}$$

Similarly, a one-form is a rank-1 tensor that we can expand in terms of basis one-forms ω^a, e.g.

$$\mathbf{C} = C_a \omega^a, \tag{A.5}$$

where the "downstairs" index a refers to a *covariant* component of $\mathbf{C}$. We say the basis one-forms ω^a are dual to the basis vectors $\mathbf{e}_b$ if

$$\omega^a \cdot \mathbf{e}_b = \delta^a{}_b, \tag{A.6}$$

where

$$\delta^a{}_b = \begin{cases} 1 & \text{if } a = b \\ 0 & \text{otherwise} \end{cases} \tag{A.7}$$

is the *Kronecker delta*. We can then compute the contraction between $\mathbf{A}$ and $\mathbf{B}$ from

$$\mathbf{A} \cdot \mathbf{B} = (A^a \mathbf{e}_a) \cdot (B_b \omega^b) = A^a B_b \, \mathbf{e}_a \cdot \omega^b = A^a B_b \, \delta_a{}^b = A^a B_a. \tag{A.8}$$

Notice that this operation does not involve the metric. Since both (A.8) and (A.4) have to hold for any tensor $\mathbf{A}$, we can compare the two and identify

$$B_a = g_{ab} B^b. \tag{A.9}$$

We refer to this operation as "lowering the index of B^a".

We define an inverse metric g^{ab} such that

$$g^{ac} g_{cb} = \delta^a{}_b. \tag{A.10}$$

We then "raise the index of B_a" using

$$B^a = g^{ab} B_b. \tag{A.11}$$

We can also show that

$$g^{ab} = \omega^a \cdot \omega^b. \tag{A.12}$$

Knowing the metric and its inverse allows us to convert a one-form into a vector and vice versa. Generally, we therefore will not distinguish between the two and instead will refer to either as a rank-1 tensor. By the same token we will not distinguish between a contraction (between a vector and a one-form) and a dot product (between two of the same kind). We find the contravariant component of a rank-1 tensor $\mathbf{A}$ by computing the dot product with the corresponding basis one-form, and we find the covariant component from the dot product with the corresponding basis vector,

$$A^a = \mathbf{A} \cdot \boldsymbol{\omega}^a \quad \text{or} \quad A_a = \mathbf{A} \cdot \mathbf{e}_a. \tag{A.13}$$

We can verify these expressions by inserting the expansions (A.1) or (A.5) for $\mathbf{A}$ and then using the duality relation (A.6).

Generalizing the concept, a rank-n tensor is an object that we can expand using a grand total of n basis vectors or one-forms. For example, a rank-2 tensor $\mathbf{T}$ can be expanded as

$$\mathbf{T} = T^a{}_b \, \mathbf{e}_a \otimes \boldsymbol{\omega}^b, \tag{A.14}$$

where $T^a{}_b$ denotes the (mixed) components of the tensor, the symbol $\otimes$ denotes the *outer product*, and the quantities $\{\mathbf{e}_a \otimes \boldsymbol{\omega}^b\}$ are the 16 spacetime *basis tensors*.

Exercise A.1 We can write the *identity tensor* $\mathbf{I}$ as $\mathbf{I} = \mathbf{e}_c \otimes \boldsymbol{\omega}^c$.
(a) Verify that $\mathbf{I} \cdot \mathbf{A} = \mathbf{A}$ for any vector $\mathbf{A}$. *Hint:* Insert (A.1) into the latter equality as well as the above expression for the identity tensor.
(b) Evaluate $I^a{}_b = \boldsymbol{\omega}^a \cdot \mathbf{I} \cdot \mathbf{e}_b$ to show that the (mixed) components of the identity tensor are given by the Kronecker delta (A.7).

We refer to a rank-2 tensor as *symmetric* if $T_{ab} = T_{ba}$ and as *antisymmetric* if $T_{ab} = -T_{ba}$. We then introduce the notation

$$T_{(ab)} = \frac{1}{2}(T_{ab} + T_{ba}) \tag{A.15}$$

for the symmetric part of a general tensor T_{ab}, and

$$T_{[ab]} = \frac{1}{2}(T_{ab} - T_{ba}) \tag{A.16}$$

for the antisymmetric part. An example of a symmetric tensor is the metric (A.3); an example of an antisymmetric tensor is the Faraday tensor (2.17) or (2.18). Both the symmetrization operator ( ) and the antisymmetrization operator [ ] can be extended to an arbitrary number of indices.

Exercise A.2 (a) Verify that the symmetrization of an antisymmetric tensor vanishes, and likewise the other way around.

(b) Show that the complete contraction between a symmetric tensor S_{ab} and an antisymmetric tensor A^{ab} vanishes,

$$S_{ab}A^{ab} = 0. \tag{A.17}$$

Throughout this book we will assume a *coordinate basis*, also known as a *holonomic* basis. We can picture a coordinate basis as one for which the basis vectors are tangent to coordinate lines, i.e. lines along which one coordinate advances while the others are held fixed. Moreover, when expressed in a coordinate basis the covariant components of the gradient of a scalar are simply the partial derivatives (i.e. the directional derivatives) of the scalar.

A common alternative to a coordinate basis is an *orthonormal* basis, usually denoted with a caret over the index, e.g. $\mathbf{e}_{\hat{a}}$. In such a basis the metric takes the form of the Minkowski metric, $g_{\hat{a}\hat{b}} = \eta_{ab}$. We can further choose coordinates so that the first derivatives of the metric at any particular point also vanish, $g_{\hat{a}\hat{b},\hat{c}} = 0$ – this is what we refer to as a *local Lorentz frame*.[2] Essentially, our ability to transform to such local Lorentz frames at any point in spacetime encodes the equivalence principle, since, when expressed in a local Lorentz frame, all physical laws will appear as they do in a flat Minkowski metric.

When expressed in an orthonormal basis, the components of the gradient of a scalar, for example, are generally not just partial derivatives with respect to the coordinates. Treatments of electrodynamics, for example, commonly adopt an orthonormal basis, which results in factors like r and $\sin\theta$ when the gradient is expressed in spherical or cylindrical coordinates.

A.2 Change of Basis

Under a change of basis, basis vectors and basis one-forms transform according to

$$\mathbf{e}_{b'} = M^a_{\ b'}\mathbf{e}_a, \tag{A.18}$$

$$\boldsymbol{\omega}^{b'} = M^{b'}_{\ a}\boldsymbol{\omega}^a, \tag{A.19}$$

---

[2] However, the second derivatives of the metric, which encode the tidal fields – the invariant measures of curvature – cannot, in general, be transformed away.

where $M^{b'}_a$ is the transformation matrix and $M^a_{b'}$ its inverse, so that

$$M^{a'}_c M^c_{b'} = \delta^{a'}_{b'}. \tag{A.20}$$

We can then use the above rules to find the transformation rules for the components of tensors, e.g.

$$A^{b'} = \mathbf{A} \cdot \boldsymbol{\omega}^{b'} = \mathbf{A} \cdot M^{b'}_a \boldsymbol{\omega}^a = M^{b'}_a \mathbf{A} \cdot \boldsymbol{\omega}^a = M^{b'}_a A^a, \tag{A.21}$$

and similarly

$$B_{b'} = M^a_{b'} A_a. \tag{A.22}$$

A specific example is a Lorentz transformation between two Lorentz frames, for which the transformation matrix $M^{a'}_b$ is often denoted by $\Lambda^{a'}_b$. For example, for a transformation into a primed reference frame that moves with speed v in the positive z-direction with respect to an unprimed reference frame, this matrix takes the form

$$M^{a'}_b = \Lambda^{a'}_b = \begin{pmatrix} \gamma & 0 & 0 & -\gamma v \\ 0 & 1 & 0 & 0 \\ 0 & 0 & 1 & 0 \\ -\gamma v & 0 & 0 & \gamma \end{pmatrix}, \tag{A.23}$$

where $\gamma \equiv (1 - v^2)^{-1/2}$ is the Lorentz factor.

Exercise A.3 (a) A particle at rest in an unprimed coordinate system has a four-velocity $u^a = (1, 0, 0, 0)$. Show that in a primed coordinate system, boosted with speed v in the negative z-direction, the particle has a four-velocity $u^{a'} = (\gamma, 0, 0, \gamma v)$, as one might expect.
(b) A particle has a four-velocity $u^a = (\gamma_1, 0, 0, \gamma_1 v_1)$ in an unprimed coordinate system. Show that in a primed coordinate system, boosted with a speed of v_2 in the negative z-direction, the particle has a four-velocity $u^{a'} = (\gamma', 0, 0, \gamma' v')$ with $v' = (v_1 + v_2)/(1 + v_1 v_2)$ – which is Einstein's famous rule for the addition of velocities. What is γ' in terms of γ_1, γ_2, and the two speeds v_1 and v_2?

Note that vectors and one-forms transform in "inverse" ways. This guarantees that the duality relation (A.6) also holds in the new reference frame,

$$\boldsymbol{\omega}^{a'} \cdot \mathbf{e}_{b'} = (M^{a'}_c \boldsymbol{\omega}^c) \cdot (M^d_{b'} \mathbf{e}_d) = M^{a'}_c M^d_{b'} (\boldsymbol{\omega}^c \cdot \mathbf{e}_d)$$
$$= M^{a'}_c M^d_{b'} \delta^c_d = M^{a'}_c M^c_{b'} = \delta^{a'}_{b'}, \tag{A.24}$$

and that the dot product (A.8) is invariant under a change of basis,

$$A^{b'} B_{b'} = M^{b'}_a A^a M^c_{b'} B_c = \delta^c_a A^a B_c = A^a B_a. \tag{A.25}$$

The dot product between two rank-1 tensors therefore transforms like a scalar, as it should.

We can generalize all the above concepts to higher-rank tensors. As we said before, a rank-n tensor can be expanded into n basis vectors or one-forms,[3] and we therefore transform the components of a rank-n tensor with n copies of the transformation matrix or its inverse. The Faraday tensor (2.17), for example, transforms as

$$F^{a'b'} = M^{a'}_{\ c} M^{b'}_{\ d} F^{cd}. \tag{A.26}$$

Exercise A.4 The electric and magnetic fields in an unprimed coordinate system are given by E^i and B^i, and the Faraday tensor by (2.17). An observer in a primed coordinate frame would identify the electric and magnetic fields from the Faraday tensor as $E^{i'} = F^{t'i'} = M^{t'}_{\ c} M^{i'}_{\ d} F^{cd}$, and similarly for the magnetic fields. Find the electric and magnetic fields as observed in a reference frame that is boosted with speed v in the positive z-direction with respect to the unprimed reference frame. *Check:* For the x'-component the answer is $E^{x'} = \gamma(E^x - vB^y)$.

For transformations between *coordinate* bases, which we will assume in the following, the transformation matrix is

$$M^{b'}_{\ a} \equiv \frac{\partial x^{b'}}{\partial x^a} = \partial_a x^{b'}. \tag{A.27}$$

As an illustration of the above concepts, consider the components of an infinitesimal displacement vector $d\mathbf{x} = dx^a\, \mathbf{e}_a$. The contravariant components dx^a measure the infinitesimal displacement between two points expressed in a coordinate system x^a. To compute the components of this vector in a different coordinate system, say a primed coordinate system $x^{b'}$, we use the chain rule to obtain

$$dx^{b'} = \frac{\partial x^{b'}}{\partial x^a} dx^a = M^{b'}_{\ a} dx^a, \tag{A.28}$$

where we have used (A.27) in the last step. As expected, the components of dx^a transform like the vector components in (A.21).

The gradient of a scalar function f, however, is most naturally expressed in terms of the covariant components, which, in a coordinate basis, are the partial derivatives of f, $\nabla f = (\partial_a f)\, \boldsymbol{\omega}^a$. Thus, the partial derivatives are components of a one-form. To transform the components

---

[3] We can always expand a so-called *mixed* tensor of rank $\binom{n}{m}$ in terms of n basis vectors and m basis one-forms. For example, $\mathbf{F} = F^a_{\ b} \mathbf{e}_a \boldsymbol{\omega}^b$ is Faraday written as a rank-$\binom{1}{1}$ tensor.

$\partial_a f$ to a new coordinate system we again use the chain rule, but this time we obtain

$$\partial_{b'} f = \frac{\partial f}{\partial x^{b'}} = \frac{\partial x^a}{\partial x^{b'}} \frac{\partial f}{\partial x^a} = M^a{}_{b'} \frac{\partial f}{\partial x^a} = M^a{}_{b'} \partial_a f \qquad (A.29)$$

as in (A.22).

To combine both concepts, consider the difference df in the values of the function f at two (nearby) points. Clearly, this difference is an invariant, i.e. it is independent of the coordinate choice. We can express this difference as the dot product between the gradient of the function and the infinitesimal displacement vector dx^a,

$$df = \nabla f \cdot d\mathbf{x} = \left(\frac{\partial f}{\partial x^a} \right) dx^a. \qquad (A.30)$$

In equation (2.38), for example, we use this relation to relate the advance of proper time as measured by a normal observer to the advance of coordinate time.

Exercise A.5 Apply the above transformation rules for the components of vectors and one-forms to show that df is indeed invariant under a coordinate transformation.

A.3 The Covariant Derivative

We now apply the above concepts to develop a derivative of tensors that also transforms like a tensor. A simple partial derivative works for derivatives of scalars, but for a higher-rank tensor the partial derivative takes into account only changes in the components of the tensor while ignoring possible changes in the basis vectors or one-forms. Our goal is to construct a *covariant derivative* that generalizes the partial derivative to higher-rank tensors.

For a tensor of rank n, say $\mathbf{A}$, a covariant derivative, denoted by $\nabla \mathbf{A}$, will be a tensor of rank $n + 1$. We can find the components of this tensor by expanding it in basis vectors and one-forms, as in Section A.1 above. If $\mathbf{A}$ is a tensor of rank-1, for example, we can find the mixed components from

$$\nabla_a A^b \equiv (\nabla \mathbf{A})_a{}^b = \mathbf{e}_a \cdot (\nabla \mathbf{A}) \cdot \boldsymbol{\omega}^b. \qquad (A.31)$$

The notation of the first term on the left-hand side is somewhat misleading, since it looks as though we are taking a derivative of the bth component of $\mathbf{A}$ in the direction of $\mathbf{e}_a$ – which would not be a tensor.

Instead, we should think of this as the mixed ab component of the tensor $\nabla \mathbf{A}$, as clarified by the middle expression above.

Next we recall that we can also expand the tensor $\mathbf{A}$ in terms of basis vectors, as in (A.1). Inserting $\mathbf{A} = A^c \mathbf{e}_c$ into (A.31) we obtain

$$\nabla_a A^b = \mathbf{e}_a \cdot \left(\nabla (A^c \mathbf{e}_c) \right) \cdot \boldsymbol{\omega}^b = \mathbf{e}_a \cdot \left((\nabla A^c) \, \mathbf{e}_c + A^c (\nabla \mathbf{e}_c) \right) \cdot \boldsymbol{\omega}^b,$$
(A.32)

where we have invoked the product (or Leibnitz) rule in the last equality (see Box A.1). The quantities A^c in these expressions are just scalar functions of the coordinates, and therefore the term ∇A^c reduces to the gradient. In a coordinate basis, the components of this gradient are the partial derivatives, $\nabla A^c = \boldsymbol{\omega}^d \partial_d A^c$, and the first term on the right-hand side of (A.32) therefore reduces to

$$\mathbf{e}_a \cdot (\nabla A^c) \, \mathbf{e}_c \cdot \boldsymbol{\omega}^b = \mathbf{e}_a \cdot \boldsymbol{\omega}^d (\partial_d A^c) \, \mathbf{e}_c \cdot \boldsymbol{\omega}^b = \delta_a{}^d (\partial_d A^c) \, \delta_c{}^b = \partial_a A^b.$$
(A.33)

We now introduce *Christoffel* symbols[4] to account for the gradients of the basis vectors in the second term on the right-hand side of (A.32),

$$\Gamma^b_{ca} \equiv (\mathbf{e}_a \cdot \nabla \mathbf{e}_c) \cdot \boldsymbol{\omega}^b = \boldsymbol{\omega}^b \cdot \nabla_a \mathbf{e}_c,$$
(A.34)

where we have used the notation $\nabla_a = \mathbf{e}_a \cdot \nabla$. Evidently, the Christoffel symbols account for changes in the basis vectors from point to point, as discussed in Section 1.2.2. Note that the components of the Christoffel symbols Γ^a_{bc} do *not* transform like a rank-3 tensor under coordinate transformations. Inserting the definition (A.34) into (A.32) now yields an expression for the covariant derivative of a rank-1 tensor,

$$\nabla_a A^b = \partial_a A^b + A^c \Gamma^b_{ca}.$$
(A.35)

We still need to figure out how to compute the Christoffel symbols, of course, but we will postpone that until we have worked out covariant derivatives of other types of tensors.

Equation (A.35) gives the covariant derivative of a rank-1 tensor expressed in terms of a contravariant component. To express a covariant derivative in terms of covariant components, we remember that the dot product $A^b B_b$ is a scalar, and therefore

---

[4] Strictly speaking, these symbols are called *connection coefficients* for a general basis, and *Christoffel* symbols only for a coordinate basis. Nevertheless, we will refer to them as Christoffel symbols throughout. We know that we have a coordinate basis if and only if the commutator between any two basis vectors vanishes, i.e. $[\mathbf{e}_a, \mathbf{e}_b] \equiv \nabla_a \mathbf{e}_b - \nabla_b \mathbf{e}_a = 0$.

$\nabla_a(A^b B_b) = \partial_a(A^b B_b)$. Demanding that covariant derivatives satisfy the product rule $\nabla_a(A^b B_b) = A^b \nabla_a B_b + B_b \nabla_a A^b$ and arguing that this must hold for an arbitrary vector A^b, shows that

$$\nabla_a B_b = \partial_a B_b - B_c \Gamma^c_{ba}. \tag{A.36}$$

Exercise A.6   Reversing the argument outlined above, show that $\nabla_a(A^b B_b) = \partial_a(A^b B_b)$ if the covariant derivatives $\nabla_a A^b$ and $\nabla_a B_b$ are given by (A.35) and (A.36).

For higher-rank tensors we add one more Christoffel symbol term for each index, with the corresponding sign for contravariant or covariant indices, e.g.

$$\nabla_a T_b{}^c = \partial_a T_b{}^c - T_d{}^c \Gamma^d_{ba} + T_b{}^d \Gamma^c_{da}. \tag{A.37}$$

So far, the above derivative operator is not yet unique. We will single out a special covariant derivative by requiring that it measure the deviation from parallel transport. If a vector **A** is parallel-transported in a direction **C** then we demand that its covariant derivative in the direction of **C** vanish, i.e. $C^c \nabla_c A^a = 0$. The same is true, of course, for a second vector **B** parallel-transported in the direction **C**. We further demand that the dot product between **A** and **B** remain unchanged while we parallel-transport the two, so that magnitudes and angles remain the same. This condition, $C^c \nabla_c(g_{ab} A^a B^b) = 0$, can only hold if the covariant derivative of the metric vanishes, implying that

$$\nabla_a g_{bc} = \partial_a g_{bc} - g_{dc} \Gamma^d_{ba} - g_{bd} \Gamma^d_{ac} = 0. \tag{A.38}$$

We refer to a covariant derivative that satisfies this condition as *compatible* or *associated* with the metric.

Exercise A.7 The *spatial covariant derivative* of the spatial metric is given by

$$D_a \gamma_{bc} = \gamma^d_a \gamma^e_b \gamma^f_c \nabla_d \gamma_{ef} \tag{A.39}$$

(see equation 2.65 and the following paragraph). Insert the definition (2.100) to show that the spatial covariant derivative is compatible with the spatial metric,

$$D_a \gamma_{bc} = 0. \tag{A.40}$$

Evidently, the Christoffel symbols are related to derivatives of the metric, as we claimed in Chapter 1. It is now a matter of finding the correct combination of partial derivatives of the metric to find an expression for a single Christoffel symbol; assuming that the Christoffel symbols are symmetric,

$$\Gamma^a_{bc} = \Gamma^a_{cb}, \tag{A.41}$$

which they are in a coordinate basis, this combination is

$$\Gamma^a_{bc} = \frac{1}{2} g^{ad} (\partial_c g_{db} + \partial_b g_{dc} - \partial_d g_{bc}). \tag{A.42}$$

Exercise A.8 (a) Use the identity (2.125), $\delta \ln g = g^{ab} \delta g_{ab}$, where g is the determinant of the metric g_{ab}, to show that[5]

$$g^{bc} \Gamma^a_{bc} = -\frac{1}{|g|^{1/2}} \partial_b \left(|g|^{1/2} g^{ab} \right). \tag{A.43}$$

(b) Use the same identity again to show that

$$\Gamma^b_{ab} = \partial_a \ln |g|^{1/2}. \tag{A.44}$$

(c) Show that that divergence of a vector can be written as

$$\nabla_a A^a = \frac{1}{|g|^{1/2}} \partial_a \left(|g|^{1/2} A^a \right). \tag{A.45}$$

Exercise A.9 Consider a flat three-dimensional space in spherical coordinates, for which the metric and inverse metric are given by

$$\gamma_{ij} = \begin{pmatrix} 1 & 0 & 0 \\ 0 & r^2 & 0 \\ 0 & 0 & r^2 \sin^2 \theta \end{pmatrix}, \quad \gamma^{ij} = \begin{pmatrix} 1 & 0 & 0 \\ 0 & r^{-2} & 0 \\ 0 & 0 & r^{-2} \sin^{-2} \theta \end{pmatrix}. \tag{A.46}$$

(a) Show that the only non-vanishing Christoffel symbols are given by

$$\begin{aligned}
\Gamma^r_{\theta\theta} &= -r, & \Gamma^r_{\varphi\varphi} &= -r \sin^2 \theta, \\
\Gamma^\theta_{\varphi\varphi} &= -\sin \theta \cos \theta, & \Gamma^\theta_{r\theta} &= r^{-1}, \\
\Gamma^\varphi_{r\varphi} &= r^{-1}, & \Gamma^\varphi_{\varphi\theta} &= \cot \theta,
\end{aligned} \tag{A.47}$$

and those that are related to the above by the symmetry $\Gamma^i_{jk} = \Gamma^i_{kj}$. *Hint:* Readers who have completed Exercise 2.24 have shown this already.

(b) Show that the Laplace operator acting on a scalar function f, defined as

$$\nabla^2 f \equiv \gamma^{ij} \nabla_i \nabla_j f, \tag{A.48}$$

reduces to the well-known expression

$$\nabla^2 f = \frac{1}{r^2} \frac{\partial}{\partial r} \left(r^2 \frac{\partial f}{\partial r} \right) + \frac{1}{r^2 \sin \theta} \frac{\partial}{\partial \theta} \left(\sin \theta \frac{\partial f}{\partial \theta} \right) + \frac{1}{r^2 \sin^2 \theta} \frac{\partial^2 f}{\partial \varphi^2}. \tag{A.49}$$

[5] We refer to Problem 7.7 in Lightman et al. (1975) for a longer list of useful identities.

Hint: You can obtain the result by using the Christoffel symbols (A.47), or, much more easily, by using the identity (A.45). Try both ways!

Finally we note that an alternative notation for $\nabla_a A^b$ is $A^b{}_{;a}$ and similarly $\nabla_a B_b = B_{b;a}$. The semicolon generalizes the comma that we sometimes use to denote a partial derivative, and reminds us to account for the change in the basis vector or one-form as well as the component in taking a covariant derivative. The terms containing Christoffel symbols account for these additional changes.

Box A.1 Some properties of the covariant derivative

The covariant derivative has all the usual properties of a gradient operator in vector calculus. We can derive these properties in a local Lorentz frame, since a tensor equation that is true in one frame is true in all frames.

For a scalar function f we have

$$\nabla_a f = \partial_a f = \frac{\partial f}{\partial x^a}, \tag{A.50}$$

where the last equality assumes a coordinate basis. The covariant derivative is linear,

$$\nabla_b(A^a + B^a) = \nabla_b A^a + \nabla_b B^a, \tag{A.51}$$

and satisfies the product (or Leibnitz) rule, e.g.

$$\nabla_a(f A^b) = f \nabla_a A^b + A^b \nabla_a f, \tag{A.52}$$

$$\nabla_a(A^b B_c) = A^b \nabla_a B_c + B_c \nabla_a A^b. \tag{A.53}$$

A vector A^a is parallel-transported along a curve with tangent vector B^b when

$$B^b \nabla_b A^a = 0. \tag{A.54}$$

For the four-dimensional covariant derivative ∇_a associated with the spacetime metric g_{bc} we have

$$\nabla_a g_{bc} = 0, \tag{A.55}$$

while, for the spatial covariant derivative D_a associated with the spatial metric γ_{bc},

$$D_a \gamma_{bc} = 0. \tag{A.56}$$

Appendix B: A Brief Introduction to Some Numerical Techniques

B.1 Functions and Derivatives

How can we solve differential equations on the computer? Before we can even get started, we need to decide how to represent functions and their derivatives. The two most common approaches adopted in numerical relativity use either *spectral* or *finite-difference* methods.[1]

In the spectral approach we write a function $f(x)$ as a linear combination of basis functions $T_k(x)$,

$$f(x) = \sum_k f_k T_k(x), \tag{B.1}$$

and our goal is to determine the coefficients f_k. Here we can determine derivatives analytically from the basis functions $T_k(x)$, e.g. trigonometric functions for periodic problems and otherwise often Chebychev polynomials. The approximation in (B.1) lies in truncating the expansion at some finite order N; for many applications this nevertheless results in a rapid exponential convergence.

In the finite-difference approach we introduce a numerical grid that covers the physical regime, say

$$x_i = x_0 + i \Delta, \tag{B.2}$$

where i is an integer and Δ is the grid *resolution* or grid spacing, which we assume here to be constant (this results in a *uniform grid*). We then express the function $f(x)$ by the values that it takes at the grid points x_i, e.g. $f_i = f(x_i)$. The approximation here lies in expressing

---

[1] An alternative approach that incorporates elements of a *discontinuous Galerkin method* is provided by, e.g., Kidder et al. (2017).

derivatives as the finite differences between neighboring values of f_i – hence the name of the approach.

Both approaches have advantages and disadvantages. As long as solutions remain smooth, spectral methods usually converge much more rapidly than finite-difference methods and are therefore faster and/or more accurate. Usually, however, they are also more cumbersome to implement, especially when there is fluid flowing and shocks are involved, whereupon the physical solutions do not remain smooth. In the interest of brevity we will limit our discussion here to finite-difference methods and will refer to the literature for introductions to alternative approaches.[2]

How can we compute derivatives from function values f_i? It is intuitive that we can compute the first derivative $f' \equiv df/dx$ at $x = x_i$, for example, from

$$f'_i = \frac{f_{i+1} - f_{i-1}}{2\Delta}, \tag{B.3}$$

which becomes exact in the limit $\Delta \to 0$. Evidently we had some options – we could also have written the first derivative as

$$f'_i = \frac{f_{i+1} - f_i}{\Delta}, \tag{B.4}$$

for example. As we will see shortly, the two expressions differ in the error that we are making. For the *centered* expression (B.3) the error decreases with Δ^2, while for the alternative *one-sided* expression it is Δ – we refer to the former as a second-order accurate finite-difference stencil, and the latter as a first-order stencil.

More systematically we can derive finite-difference expressions using Taylor expansions about the point x_i. Using the immediate neighbors x_{i+1} and x_{i-1} only, we have

$$f_{i+1} = f_i + \Delta f'_i + \frac{1}{2}\Delta^2 f''_i + \mathcal{O}(\Delta^3),$$
$$f_i = f_i, \tag{B.5}$$
$$f_{i-1} = f_i - \Delta f'_i + \frac{1}{2}\Delta^2 f''_i + \mathcal{O}(\Delta^3).$$

Subtracting the first and last of these equations, and dividing by 2Δ, we obtain (B.3). We also see that, unlike in the case of the one-sided scheme, the f'' term cancels out so that the leading-order error term

---

[2] An excellent starting point for all things numerical is Press et al. (2007).

scales with Δ^2 rather than Δ. In general, we say that a numerical scheme is nth-order accurate, or converges to nth order, if the error scales with Δ^n.

Exercise B.1 Show that a one-sided second-order expression for the first derivative is given by

$$f_i' = \frac{3f_i - 4f_{i-1} + f_{i-2}}{2\Delta}. \tag{B.6}$$

More elegantly, though, we can write equations (B.5) as the matrix equation

$$\mathbf{f} = \mathbf{M} \cdot \mathbf{d} + \mathbf{e}. \tag{B.7}$$

Here $\mathbf{f} = (f_{i+1}, f_i, f_{i-1})$ is a vector containing the function values, $\mathbf{d} = (f_i, f_i', f_i'')$ is a vector containing the derivatives at the grid point x_i, and $\mathbf{M}$ is the matrix

$$\mathbf{M} = \begin{pmatrix} 1 & \Delta & \Delta^2/2 \\ 1 & 0 & 0 \\ 1 & -\Delta & \Delta^2/2 \end{pmatrix}. \tag{B.8}$$

Expanding (B.5) to one additional order we can write the error term explicitly as

$$\mathbf{e} = (\Delta^3 f_i^{(3)}/6, \ 0, \ -\Delta^3 f_i^{(3)}/6), \tag{B.9}$$

where $f_i^{(n)}$ denotes the nth derivative of the function f at the grid point x_i. Solving (B.7) for $\mathbf{d}$ we have

$$\mathbf{d} = \mathbf{M}^{-1} \cdot \mathbf{f} - \mathbf{M}^{-1} \cdot \mathbf{e}, \tag{B.10}$$

where the inverse matrix $\mathbf{M}^{-1}$ is given by

$$\mathbf{M}^{-1} = \begin{pmatrix} 0 & 1 & 0 \\ 1/(2\Delta) & 0 & -1/(2\Delta) \\ 1/\Delta^2 & -2/\Delta^2 & 1/\Delta^2 \end{pmatrix}. \tag{B.11}$$

Not surprisingly, the first line of (B.10) yields the identity $f_i = f_i$, which is not particularly interesting. From the second line we obtain

$$f_i' = \frac{f_{i+1} - f_{i-1}}{2\Delta} + \mathcal{O}(\Delta^2), \tag{B.12}$$

which is identical to (B.3) except that we have now included the scaling of the error, as obtained from $\mathbf{M}^{-1} \cdot \mathbf{e}$. The third line in (B.10) provides us with an expression for the second derivative:

$$f_i'' = \frac{f_{i+1} - 2f_i + f_{i-1}}{\Delta^2} + \mathcal{O}(\Delta^2). \qquad (B.13)$$

Note that the leading-order error term, as computed from $\mathbf{M}^{-1} \cdot \mathbf{e}$, cancels; therefore this expression is still second-order accurate even though we have divided by Δ^2 in $\mathbf{M}^{-1}$. We have included the explicit error terms in $\mathbf{e}$ rather than just their scaling, so that this cancelation can be seen more easily.

> **Exercise B.2** Extend the expressions (B.5) to also include f_{i+2} and f_{i-2}, and expand up to order Δ^5, i.e. including up to fifth derivatives. Write the equations in the matrix form (B.7), where the fifth derivatives enter the error terms only, invert the new 5×5 matrix $\mathbf{M}$, and obtain new expressions for the first and second derivatives. Show that these expressions are fourth-order accurate, meaning that the error decreases with Δ^4.

As a simple sanity check, we observe that all derivatives must vanish for a constant function. That means that the coefficients in finite-difference stencils for any derivative must always add up to zero. Evidently, this is the case for both (B.12) and (B.13), and it is also true for the stencils found in Exercises B.1 and B.2.

We distinguish between *cell-centered* and *vertex-centered* grids, the difference being the following. Imagine we want to solve a problem in a certain physical domain, say from $x = x_{min}$ to $x = x_{max}$, with boundary conditions given at both boundaries. We then divide this physical domain into, say, N *grid cells*. Assuming a uniform grid, each grid cell has a width $\Delta = (x_{max} - x_{min})/N$, and the first grid cell extends from x_{min} to $x_{min} + \Delta$. We now have a choice of where to put our grid points. As the name suggests, we place the *grid points* at the centers of the grid cells for a cell-centered grid, while for a vertex-centered grid we place them at the vertices.[3] For the former we then end up with N grid points located at

$$x_i = x_{min} + (i + 1/2)\Delta, \quad i = 0, \ldots, N - 1 \quad \text{(cell-centered)}, \qquad (B.14)$$

while for the latter we have $N + 1$ grid points located at

$$x_i = x_{min} + i\Delta, \quad i = 0, \ldots, N \quad \text{(vertex-centered)}. \qquad (B.15)$$

Choosing one or the other affects how the boundary conditions are imposed but not the finite-differencing in the interior of the grid.

---

[3] So-called *staggered* grids, which used to be more common than they are now, use a combination of both, namely a cell-centered grid for some variables and a vertex-centered grid for others.

Different types of boundary conditions are commonly encountered. *Dirichlet* boundary conditions set the dependent variable to certain given function values on the boundary, while *Neumann* boundary conditions set the derivative of the function to given values. We encounter examples of these when we impose symmetries. Say the solution features a symmetry about $x = 0$; then an antisymmetric function must have $f(0) = 0$ (a Dirichlet boundary condition), while a symmetric function must have $f'(0) = 0$ (a Neumann boundary condition). Another common boundary condition is the *Robin* boundary condition, which forces the solution to drop off with inverse distance from the origin, i.e. $rf = const$ as $r \rightarrow \infty$. One way to impose this condition is to set the derivative of rf to zero on the boundary, or to set rf on the boundary equal to its value just inside the boundary. Finally, a common boundary condition for dynamical problems is an *outgoing-wave* boundary condition, which allows outgoing waves to leave the numerical grid but aims to suppress any waves from entering – more on that in Section B.3.1 below.

Now we are ready to discuss how to use the above expressions in the solution of partial differential equations. From a computational perspective we distinguish boundary-value problems and initial-value problems. As the name suggests, boundary-value problems require boundary conditions only; these problems arise for elliptic equations, for example the Poisson equation. Initial-value problems, however, require initial data also; these problems arise for parabolic equations, for example the diffusion equation, and hyperbolic equations, for example the wave equation. In the context of numerical relativity we encounter boundary-value problems when solving the constraint equations of Chapter 3, as well as some coordinate conditions (for example the maximal slicing condition 4.36), while we are dealing with initial-value problems when we solve the evolution equations of Chapter 4 as well as other coordinate conditions (for example the moving-puncture gauge conditions 4.46–4.48).

B.2 Boundary-Value Problems

B.2.1 Linear Problems

To start with a concrete example, assume we want to solve the linear one-dimensional problem

$$\partial_x^2 f + gf = s \tag{B.16}$$

subject to Robin boundary conditions at the outer boundaries $x = \pm x_{OB}$. In (B.16) the functions $g = g(x)$ and $s = s(x)$ are assumed to be given as a function of the coordinates, and our goal is to find $f = f(x)$. First we set up a numerical grid; say we choose a cell-centered grid (B.14) with $x_{min} = -x_{OB}$ and $x_{max} = x_{OB}$. On the boundaries we then implement the Robin boundary conditions as

$$x_0 f_0 - x_1 f_1 = 0 \tag{B.17}$$

and

$$x_{N-1} f_{N-1} - x_{N-2} f_{N-2} = 0, \tag{B.18}$$

and in the interior we adopt the second-order differencing stencil (B.13) to obtain

$$f_{i+1} + (\Delta^2 g_i - 2) f_i + f_{i-1} = \Delta^2 s_i, \quad 0 < i < N - 1 \tag{B.19}$$

(where $g_i = g(x_i)$ and $s_i = s(x_i)$). The above equations now form a system of linear equations for the f_i, which we can organize as the matrix equation

$$\mathbf{A} \cdot \mathbf{f} = \mathbf{s}. \tag{B.20}$$

Here the matrix $\mathbf{A}$ is given by

$$\mathbf{A} = \begin{pmatrix} x_0 & -x_1 & & & & \\ 1 & \Delta^2 g_1 - 2 & 1 & & & \\ & \ddots & \ddots & \ddots & & \\ & & 1 & \Delta^2 g_{N-2} - 2 & 1 \\ & & & -x_{N-2} & x_{N-1} \end{pmatrix}, \tag{B.21}$$

the solution vector $\mathbf{f}$ by

$$\mathbf{f} = \begin{pmatrix} f_0 \\ \vdots \\ f_{N-1} \end{pmatrix}, \tag{B.22}$$

and the right-hand side by

$$\mathbf{s} = \begin{pmatrix} 0 \\ \Delta^2 s_1 \\ \vdots \\ \Delta^2 s_{N-2} \\ 0 \end{pmatrix}. \tag{B.23}$$

We can now use standard numerical techniques to invert the matrix $\mathbf{A}$ and find the solution from

$$\mathbf{f} = \mathbf{A}^{-1} \cdot \mathbf{s}. \tag{B.24}$$

Even better: we could recognize that the matrix $\mathbf{A}$ is *tridiagonal*, meaning that it has non-zero entries on the diagonal and its immediate neighbors only; we can invert such a matrix much more quickly than a general matrix.[4] In either case, the problem is solved.

In higher dimensions things get a little more complicated. Say we now want to solve the three-dimensional generalization of equation (B.16),

$$D^2 f + gf = s, \tag{B.25}$$

where D^2 is the Laplace operator. For a start, we will now have to introduce a grid for each dimension. For simplicity we will assume Cartesian coordinates and that both the grid spacing Δ and the number of grid points N are the same in every dimension. Grid functions will now have three indices, e.g. $f_{ijk} = f(x_i, y_j, z_k)$. Using expressions like (B.13) for every dimension we can finite-difference the Laplace operator according to

$$\begin{aligned}
(D^2 f)_{ijk} &= (\partial_x^2 f)_{ijk} + (\partial_y^2 f)_{ijk} + (\partial_z^2 f)_{ijk} \\
&= \frac{1}{\Delta^2} (f_{i+1,j,k} + f_{i-1,j,k} + f_{i,j+1,k} + f_{i,j-1,k} \\
&\quad + f_{i,j,k+1} + f_{i,j,k-1} - 6 f_{ijk}),
\end{aligned} \tag{B.26}$$

and the generalization of the interior-grid expression (B.19) becomes

$$\begin{aligned}
f_{i+1,j,k} + f_{i,j+1,k} + f_{i,j,k+1} + (\Delta^2 g_{ijk} - 6) f_{ijk} + f_{i-1,j,k} \\
+ f_{i,j-1,k} + f_{i,j,k-1} = \Delta^2 s_{ijk}
\end{aligned} \tag{B.27}$$

for $0 < i, j, k < N-1$ (we will separate the different indices by commas when it makes the notation clearer, but not otherwise).

A priori, we cannot accommodate the f_{ijk} into a single solution vector as in (B.22) and so we cannot easily write (B.27) as a matrix equation. In order to do that, we need to organize the f_{ijk} into a single one-dimensional vector $\mathbf{F}$ of length N^3. We label the components F_I of this vector with a *super-index* I that runs over all combinations of the indices i, j, and k. A possible choice for this super-index is

$$I = i + Nj + N^2 k. \tag{B.28}$$

---

[4] See, e.g., Press et al. (2007).

If we label the point (i, j, k) as I then the point $(i + 1, j, k)$ will correspond to $I + 1$, and $(i, j - 1, k)$ to $I - N$, etc. We can therefore rewrite (B.27) in terms of F_I as

$$F_{I+N^2} + F_{I+N} + F_{I+1} + (\Delta^2 G_I - 6)F_I + F_{I-1} + F_{I-N} + F_{I-N^2}$$
$$= \Delta^2 S_I, \quad \text{(B.29)}$$

where $G_I = g_{ijk}$ and $S_I = s_{ijk}$. Equation (B.29) applies to values of I corresponding to interior grid points $0 < i, j, k < N - 1$ only; on the boundaries we again implement Robin boundary conditions, e.g.

$$r_{ijk}F_I - r_{i,j,k+1}F_{I+N^2} = 0 \quad \text{(B.30)}$$

for a point on the lower z-boundary, where $k = 0$.

Equation (B.29) together with the boundary conditions can now be cast as a matrix equation, similar to (B.20), for F_I. We notice that the matrix $\mathbf{A}$ is no longer tridiagonal, since it now has non-zero elements not only on the diagonal and its immediate neighbors but also N elements away and N^2 elements away. We call this structure *band-diagonal*; in this case the matrix has seven bands, namely the diagonal as well as the bands removed by N^2, N, and one element on either side from the diagonal.

Also notice that the matrix $\mathbf{A}$ becomes *very* large *very* quickly. For a grid with N grid points in each dimension, the solution vector $\mathbf{F}$ has N^3 elements in three dimensions, and the matrix $\mathbf{A}$ then has N^6 elements. For quite a modest grid, with $N = 100$, say, the matrix $\mathbf{A}$ has 10^{12} (!) elements. This points to a general challenge when dealing with boundary-value problems: efficiency. Of course, only about 7×10^6 of the 10^{12} elements in this example will be non-zero. It is therefore important to take advantage of the band-diagonal structure of the matrix. A number of specialized software packages make it quite easy to invert such matrices even in a parallel-computing environment.[5]

B.2.2 Nonlinear Problems

The methods of Section B.2.1 make it possible to solve linear boundary-value problems, but what if our equations also involve nonlinear terms – like equations (3.45) or (3.48), say? In this case we can reduce the problem to a linear problem using *linearization* followed by *iteration*.

---

[5] Examples include the LAPACK library (see www.netlib.org/lapack/), PETSc (see www.mcs .anl.gov/petsc/) and Trilinos (see https://trilinos.github.io).

Consider a problem of the form

$$D^2 f = h(f), \tag{B.31}$$

where h is some potentially nonlinear function of f (see Section B.2.3 for a concrete example). Write

$$f^{[n+1]} = f^{[n]} + \delta f, \tag{B.32}$$

where the superscript $[n]$ denotes the nth iteration. Assuming $\delta f \ll f^{[n]}$ we can expand, to linear order, the $(n+1)$th iteration for h about the (previous) nth iteration,

$$h(f^{[n+1]}) = h(f^{[n]} + \delta f) = h(f^{[n]}) + (\delta f)h'(f^{[n]}) + \mathcal{O}\left((\delta f)^2\right), \tag{B.33}$$

where $h' = dh/df$. Inserting $f^{[n+1]}$ for f into (B.31) we then obtain

$$D^2(\delta f) - h'(f^{[n]})\,(\delta f) = -D^2 f^{[n]} + h(f^{[n]}). \tag{B.34}$$

We now define the *residual* of the nth iteration as the difference between the left- and right-hand sides of our equation (B.31),

$$\mathcal{R}^{[n]} = D^2 f^{[n]} - h(f^{[n]}) \tag{B.35}$$

in terms of which equation (B.34) becomes

$$D^2(\delta f) - h'(f^{[n]})\,(\delta f) = -\mathcal{R}^{[n]}. \tag{B.36}$$

But this equation is an equation of the form (B.25) for δf, with $g = -h'(f^{[n]})$ and $s = -\mathcal{R}^{[n]}$. We can solve this equation repeatedly, starting with some initial guess, $f^{[0]}$. For each iteration step we compute the residual $\mathcal{R}^{[n]}$ and $h'(f^{[n]})$ from our current guess $f^{[n]}$ (i.e. we have to update both the right-hand side s and the matrix A in our matrix equation), and then solve (B.36) to find δf, and finally find the next guess $f^{[n+1]}$ from (B.32). If all goes well, our iteration converges, and we can terminate the iteration when the norm of the residual $\mathcal{R}$ drops below a given tolerance. Assuming that the operator on the left-hand side of (B.36) is sufficiently well behaved, we see that the corrections δf will become smaller as the residuals $\mathcal{R}$ decrease.

B.2.3 A Worked Example: Puncture Initial Data

To see how the techniques of the previous section work for a concrete example, consider equation (3.48),

$$\bar{D}^2 u = -\beta \left(\alpha + \alpha u + 1 \right)^{-7}, \tag{B.37}$$

where u is the correction of the Schwarzschild conformal factor for Bowen–York black holes carrying either angular or linear momentum (see equation 3.46), and where the functions α and β are given by (3.47) and (3.49). Comparing with (B.31) we identify f with u and obtain

$$h(f) = h(u) = -\beta(\alpha + \alpha u + 1)^{-7}. \tag{B.38}$$

Evidently equation (B.37) is nonlinear in u, and we will therefore use an iteration $u^{[n+1]} = u^{[n]} + \delta u$ as in Section B.2.2, i.e. we have to solve (B.36) with

$$h'(u^{[n]}) = 7\alpha\beta \left(\alpha + \alpha u^{[n]} + 1 \right)^{-8}. \tag{B.39}$$

We will now treat the problem for a single black hole.

Note that physical dimensions enter the problem only through the puncture mass $\mathcal{M}$ in the definition of α, equation (3.47). We may therefore define non-dimensional variables by multiplying all dimensional quantities with appropriate powers of $\mathcal{M}$, e.g.

$$\bar{x}^i = x^i / \mathcal{M}, \quad \bar{P}^i = P^i / \mathcal{M}, \tag{B.40}$$

in which case $\mathcal{M}$ drops out of the problem completely (recall the geometrized units that we introduced in Section 1.2.1). In practice we achieve exactly the same effect, of course, by setting $\mathcal{M}$ to unity. We now write our entire code in terms of the non-dimensional variables, which carry overbars, and thereby avoid having to commit to a specific-black-hole mass. Should we be interested in the results for a specific mass, we can always recover dimensional quantities from (B.40).

Now we will throw caution to the wind – we will ignore the warnings concerning efficiency at the end of Section B.2.1 and will, in the interests of clarity and simplicity, implement a "brute force" routine to solve the above problem. We list the Python script `puncture.py` below, but the code can also be downloaded from www.cambridge.org/NRStartingfromScratch. It implements the above methods using a hierarchy of classes: (a) a class `EllipticSolver` provides an interface to a `scipy.linalg` matrix solver to solve Poisson-type elliptic equations subject to a Robin boundary condition;

(b) this class is instantiated in a class `Puncture` that solves the equations of the puncture method; and finally (c) the `main` routine makes calls to `Puncture` to construct black-hole initial data for a number of different parameter choices.

Before we list the script, we have a disclaimer: coding is always an exercise in compromise, in which a number of different goals, including efficiency, readability, and generalizability, are weighed against each other. In the script here, as well as the script in Section B.3.2, we have not paid much attention to efficiency – presumably, somebody interested in a large-scale highly efficient code would not use Python anyway. Instead we have tried to make the scripts as readable as possible. We hope that the reader can easily identify equations in the text (we have even coded some operators, rather than using Python libraries, so that the reader can see these operators "at work"), can run and modify the scripts (see the exercises below for suggestions), and can get a sense of how the algorithms described above can be implemented.

```python
"""Code to construct puncture initial data for single black hole."""
import sys
from numpy import zeros, size, sqrt, linspace
import scipy.linalg as la


class EllipticSolver:
    """Class Elliptic solves Poisson-type elliptic equations of the form:
            D^2 sol + fct sol = rhs
    where
            - D^2 is the flat Laplace operator
            - fct and rhs are user-supplied functions of the coordinates x, y, z,
            - and sol is the solution.

    To use this class:
            - initialize the class, providing Cartesian coordinates x, y, and z
            - call setup_matrix(fct) to set up the operator
            - call setup_rhs(rhs) to set up the right-hand side
            - then a call to solve() returns the solution sol
    """

    def __init__(self, x, y, z):
        """Constructor - provide Cartesian coordinates, all of length n_grid,
        as arguments.
        """

        print(" Setting up Poisson solver...")
        self.n_grid = size(x)
        self.delta = x[1] - x[0]

        # set up storage for matrix, solution, r.h.s.
        # Note: "sol" and "rhs" will store functions in 3d format, while
        # "sol_1d" and "rhs_1d" will store functions in 1d format using
        # super-index
        nnn = self.n_grid ** 3
        self.rhs_1d = zeros(nnn)
        self.A = zeros((nnn, nnn))
        self.sol = zeros((self.n_grid, self.n_grid, self.n_grid))
        self.rad = zeros((self.n_grid, self.n_grid, self.n_grid))
```

```python
        # compute radius
        for i in range(0, self.n_grid):
            for j in range(0, self.n_grid):
                for k in range(0, self.n_grid):
                    rad2 = x[i] ** 2 + y[j] ** 2 + z[k] ** 2
                    self.rad[i, j, k] = sqrt(rad2)


    def setup_matrix(self, fct):
        """Set up matrix A."""

        n_grid = self.n_grid

        # Use Robin boundary conditions (B.30) to set up boundaries
        i = 0  # lower x-boundary
        for j in range(0, n_grid):
            for k in range(0, n_grid):
                index = self.super_index(i, j, k)
                self.A[index, index] = self.rad[i, j, k]
                self.A[index, index + 1] = -self.rad[i + 1, j, k]

        i = n_grid - 1  # upper x-boundary
        for j in range(0, n_grid):
            for k in range(0, n_grid):
                index = self.super_index(i, j, k)
                self.A[index, index] = self.rad[i, j, k]
                self.A[index, index - 1] = -self.rad[i - 1, j, k]

        j = 0  # lower y-boundary
        for i in range(1, n_grid - 1):
            for k in range(0, n_grid):
                index = self.super_index(i, j, k)
                self.A[index, index] = self.rad[i, j, k]
                self.A[index, index + n_grid] = -self.rad[i, j + 1, k]

        j = n_grid - 1  # upper y-boundary
        for i in range(1, n_grid - 1):
            for k in range(0, n_grid):
                index = self.super_index(i, j, k)
                self.A[index, index] = self.rad[i, j, k]
                self.A[index, index - n_grid] = -self.rad[i, j - 1, k]

        k = 0  # lower z-boundary
        for i in range(1, n_grid - 1):
            for j in range(1, n_grid - 1):
                index = self.super_index(i, j, k)
                self.A[index, index] = self.rad[i, j, k]
                self.A[index, index + n_grid * n_grid] = -self.rad[i, j, k + 1]

        k = n_grid - 1  # upper z-boundary
        for i in range(1, n_grid - 1):
            for j in range(1, n_grid - 1):
                index = self.super_index(i, j, k)
                self.A[index, index] = self.rad[i, j, k]
                self.A[index, index - n_grid * n_grid] = -self.rad[i, j, k - 1]

        # use (B.29) to fill matrix in interior
        for i in range(1, n_grid - 1):
            for j in range(1, n_grid - 1):
                for k in range(1, n_grid - 1):
                    index = self.super_index(i, j, k)

                    # diagonal element
                    self.A[index, index] = -6. + self.delta ** 2 * fct[i, j, k]

                    # off-diagonal elements
                    self.A[index, index - 1] = 1.0
                    self.A[index, index + 1] = 1.0
```

```
                    self.A[index, index - n_grid] = 1.0
                    self.A[index, index + n_grid] = 1.0

                    self.A[index, index - n_grid * n_grid] = 1.0
                    self.A[index, index + n_grid * n_grid] = 1.0

    def setup_rhs(self, rhs):
        """Setup right-hand side of matrix equation"""

        n_grid = self.n_grid
        for i in range(1, n_grid - 1):
            for j in range(1, n_grid - 1):
                for k in range(1, n_grid - 1):
                    index = self.super_index(i, j, k)
                    self.rhs_1d[index] = self.delta ** 2 * rhs[i, j, k]

    def solve(self):
        """Interface to scipy.linalg matrix solver,
        returns sol (in 3d format)."""

        # solve matrix using scipy.linalg interface...
        sol_1d = la.solve(self.A, self.rhs_1d)

        # ... then translate from superindex to 3d
        for i in range(0, self.n_grid):
            for j in range(0, self.n_grid):
                for k in range(0, self.n_grid):
                    index = self.super_index(i, j, k)
                    self.sol[i, j, k] = sol_1d[index]

        return self.sol

    def super_index(self, i, j, k):
        """Compute super index, see (B.28)."""
        return i + self.n_grid * (j + self.n_grid * k)


class Puncture:
    """Class that handles construction of puncture data.

    To use this class,
        - initialize class with physical parameters as arguments
        - then call construct_solution.
    """

    def __init__(self, bh_loc, lin_mom, n_grid, x_out):
        """Arguments to constructor specify physical parameters:
        - location of puncture (bh_loc)
        - linear momentum (lin_mom)
        - size of grid (n_grid)
        - outer boundary (x_out).
        """
        self.bh_loc = bh_loc
        self.lin_mom = lin_mom
        # echo out parameters
        print(" Constructing class Puncture for single black hole")
        print("    at bh_loc = (", bh_loc[0], ",", bh_loc[1], ",",
                bh_loc[2], ")")
        print("    with momentum p = (", lin_mom[0], ",",
                lin_mom[1], ",", lin_mom[2], ")")
        print(" Using", n_grid, "\b^3 gridpoints with outer boundary at", x_out)
        # set up grid
        self.n_grid = n_grid
        self.x_out = x_out
```

```
        self.delta = 2.0 * x_out / n_grid


        # set up coordinates: use cell-centered grid covering (-x_out, x_out)
        # in each dimension; see (B.14)
        half_delta = self.delta / 2.0
        self.x = linspace(half_delta - x_out, x_out -
                          half_delta, n_grid)
        self.y = linspace(half_delta - x_out, x_out -
                          half_delta, n_grid)
        self.z = linspace(half_delta - x_out, x_out -
                          half_delta, n_grid)

        # allocate elliptic solver
        self.solver = EllipticSolver(self.x, self.y, self.z)

        # allocate memory for functions u, alpha, beta, and residual
        self.alpha = zeros((n_grid, n_grid, n_grid))
        self.beta = zeros((n_grid, n_grid, n_grid))
        self.u = zeros((n_grid, n_grid, n_grid))
        self.res = zeros((n_grid, n_grid, n_grid))

    def construct_solution(self, tol, it_max):
        """Construct solution iteratively, provide tolerance and maximum
        number of iterations as arguments."""

        self.setup_alpha_beta()
        residual_norm = self.residual()
        print(" Initial Residual = ", residual_norm)
        print(" Using up to", it_max, "iteration steps to reach tolerance of",
              tol)

        # now iterate...
        it_step = 0
        while residual_norm > tol and it_step < it_max:
            it_step += 1
            self.update_u()
            residual_norm = self.residual()
            print(" Residual after", it_step, "iterations :", residual_norm)
        if (residual_norm < tol):
            print(" Done!")
        else:
            print(" Giving up...")


    def update_u(self):
        """Function that updates u using Poisson solver;
        takes one iteration step.
        """

        # set up linear term and right-hand side for SolvePoisson...
        n_grid = self.n_grid
        fct = zeros((n_grid, n_grid, n_grid))
        rhs = zeros((n_grid, n_grid, n_grid))

        for i in range(1, n_grid - 1):
            for j in range(1, n_grid - 1):
                for k in range(1, n_grid - 1):
                    # compute h' from (B.39)
                    temp = self.alpha[i, j, k] * (1.0 + self.u[i, j, k]) + 1.0
                    fct[i, j, k] = (-7.0 * self.beta[i, j, k] *
                                    self.alpha[i, j, k] / temp ** 8)
                    rhs[i, j, k] = -self.res[i, j, k]

        # now update Poisson solver
        self.solver.setup_matrix(fct)
```

```
      # set up right-hand side
      self.solver.setup_rhs(rhs)

      # solve to find delta_u, see (B.36)
      delta_u = self.solver.solve()

      # update u
      self.u += delta_u


def residual(self):
    """Evaluate residual, see (B.35)."""

    residual_norm = 0.0
    for i in range(1, self.n_grid - 1):
        for j in range(1, self.n_grid - 1):
            for k in range(1, self.n_grid - 1):

                # compute left-hand side: Laplace operator
                ddx = (self.u[i + 1, j, k] - 2.0 * self.u[i, j, k] +
                       self.u[i - 1, j, k])
                ddy = (self.u[i, j + 1, k] - 2.0 * self.u[i, j, k] +
                       self.u[i, j - 1, k])
                ddz = (self.u[i, j, k + 1] - 2.0 * self.u[i, j, k] +
                       self.u[i, j, k - 1])
                lhs = (ddx + ddy + ddz) / self.delta ** 2

                # compute right-hand side,
                # recall h = - beta/(alpha + alpha u + 1)^7
                temp = self.alpha[i, j, k] * (1.0 + self.u[i, j, k]) + 1.0
                rhs = -self.beta[i, j, k] / temp ** 7

                # then compute difference to get residual, see (B.35)
                self.res[i, j, k] = lhs - rhs
                residual_norm += self.res[i, j, k] ** 2

    residual_norm = sqrt(residual_norm) * self.delta ** 3
    return residual_norm


def setup_alpha_beta(self):
    """Set up functions alpha and beta."""

    n_grid = self.n_grid
    p_x = self.lin_mom[0]
    p_y = self.lin_mom[1]
    p_z = self.lin_mom[2]

    for i in range(0, n_grid):
        for j in range(0, n_grid):
            for k in range(0, n_grid):
                s_x = self.x[i] - self.bh_loc[0]
                s_y = self.y[j] - self.bh_loc[1]
                s_z = self.z[k] - self.bh_loc[2]
                s2 = s_x ** 2 + s_y ** 2 + s_z ** 2
                s_bh = sqrt(s2)
                l_x = s_x / s_bh
                l_y = s_y / s_bh
                l_z = s_z / s_bh
                lP = l_x * p_x + l_y * p_y + l_z * p_z

                # construct extrinsic curvature, see (3.43)
                fac = 3.0 / (2.0 * s2)
                A_xx = fac * (2.0 * p_x * l_x - (1.0 - l_x * l_x) * lP)
                A_yy = fac * (2.0 * p_y * l_y - (1.0 - l_y * l_y) * lP)
                A_zz = fac * (2.0 * p_z * l_z - (1.0 - l_z * l_z) * lP)
                A_xy = fac * (p_x * l_y + p_y * l_x + l_x * l_y * lP)
                A_xz = fac * (p_x * l_z + p_z * l_x + l_x * l_z * lP)
```

```python
                        A_yz = fac * (p_y * l_z + p_z * l_y + l_y * l_z * 1P)


                        # compute A_{ij} A^{ij}
                        A2 = (
                            A_xx ** 2 + A_yy ** 2 + A_zz ** 2 +
                            2.0*(A_xy ** 2 + A_xz ** 2 + A_yz ** 2)
                            )

                        # now compute alpha and beta from (3.47) and (3.49)
                        self.alpha[i, j, k] = 2.0 * s_bh
                        self.beta[i, j, k] = self.alpha[i, j, k] ** 7 * A2 / 8.0


    def write_to_file(self):
        """Function that writes solution to file."""

        filename = "Puncture_" + str(self.n_grid) + "_" + str(self.x_out)
        filename = filename + ".data"
        out = open(filename, "w")
        if out:
            k = self.n_grid // 2
            out.write(
                "# Data for black hole at x = (%f,%f,%f)\n"
                % (self.bh_loc[0], self.bh_loc[1], self.bh_loc[2])
                )
            out.write("# with linear momentum P = (%f, %f, %f)\n" %
                        (self.lin_mom))
            out.write("# in plane for z = %e \n" % (self.z[k]))
            out.write("# x            y            u            \n")
            out.write("#==========================================\n")
            for i in range(0, self.n_grid):
                for j in range(0, self.n_grid):
                    out.write("%e  %e  %e\n" % (self.x[i], self.y[j],
                                                self.u[i, j, k]))
                out.write("\n")
            out.close()
        else:
            print(" Could not open file", filename,"in write_to_file()")
            print(" Check permissions?")
#
#=====================================================================
# Main routine: defines parameters, sets up puncture solver, and
# then finds solution
#=====================================================================
#
def main():
    """Main routine..."""
    print(" ----------------------------------------------------")
    print(" --- puncture.py --- use flag -h for list of options ---")
    print(" ----------------------------------------------------")
    #
    # set default values for variables
    #
    # location of black hole:
    loc_x = 0.0
    loc_y = 0.0
    loc_z = 0.0
    # momentum of black hole:
    p_x = 1.0
    p_y = 0.0
    p_z = 0.0
    # number of grid points
    n_grid = 16
    # location of outer boundary
    x_out = 4.0
    # tolerance and maximum number of iterations
    tol = 1.0e-12
```

```
    it_max = 50
    #
    # now look for flags to overwrite default values
    #
    for i in range(len(sys.argv)):
        if sys.argv[i] == "-h":
            usage()
            return
        if sys.argv[i] == "-n_grid":
            n_grid = int(sys.argv[i+1])
        if sys.argv[i] == "-x_out":
            x_out = float(sys.argv[i+1])
        if sys.argv[i] == "-loc_x":
            loc_x = float(sys.argv[i+1])
        if sys.argv[i] == "-loc_y":
            loc_y = float(sys.argv[i+1])
        if sys.argv[i] == "-loc_z":
            loc_z = float(sys.argv[i+1])
        if sys.argv[i] == "-p_x":
            p_x = float(sys.argv[i+1])
        if sys.argv[i] == "-p_y":
            p_y = float(sys.argv[i+1])
        if sys.argv[i] == "-p_z":
            p_z = float(sys.argv[i+1])
        if sys.argv[i] == "-tol":
            tol = float(sys.argv[i+1])
        if sys.argv[i] == "-it_max":
            it_max = int(sys.argv[i+1])

    # location of puncture
    bh_loc = ( loc_x, loc_y, loc_z )
    # linear momentum
    lin_mom = ( p_x, p_y, p_z )
    #
    # set up Puncture solver
    black_hole = Puncture(bh_loc, lin_mom, n_grid, x_out)
    #
    # and construct solution
    black_hole.construct_solution(tol, it_max)
    #
    # and write results to file
    black_hole.write_to_file()


def usage():
    print("Constructs puncture initial data for single black hole.")
    print("")
    print("The following options can be used to overwrite default parameters")
    print("\t-n_grid: number of grid points [default: 16]")
    print("\t-x_out: location of outer boundary [4.0]")
    print("\t-loc_x, -loc_y, -loc_z: location of black hole [(0.0, 0.0, 0.0)]")
    print("\t-p_x, -p_y, -p_z: lin. momentum of black hole [(1.0, 0.0, 0.0)]")
    print("\t-tol: tolerance for elliptic solver [1.e-12]")
    print("\t-it_max: maximum number of iterations [50]")
    print("For example, to construct data with x_out = 6.0, call")
    print("\tpython3 puncture.py -x_out 6.0")

if __name__ == '__main__':
    main()
```

Downloading the script `puncture.py` from www.cambridge.org/NRStarting fromScratch, it can be run at command line with

```
python3 puncture.py
```

This should produce puncture initial data for a single black hole with certain default parameters, including n_grid = 16 and x_out = 4.0. Running the script with the option -h, i.e.

```
python3 puncture.py -h
```

will return a list of all parameters, their default values, and flags that can be used to overwrite these default values. For example, to produce data with n_grid = 24 and x_out = 6.0, call

```
python3 puncture.py -n_grid 24 -x_out 6.0
```

An example for a black hole with linear momentum $\bar{P}^i = (1,0,0)$ is shown in Fig. 3.1. The image was generated with the script puncture_plot.py, which can also be downloaded from www.cambridge.org/NRStartingfromScratch.

All the above scripts were written for Python3 but should run with some earlier versions of Python also. Depending on the computer platform, and on which libraries have been installed (in particular matplotlib, which is used in puncture_plot.py), readers may want to try different versions of Python. Of course, readers could also upload these routines into a jupyter notebook or their favorite IDE.

> **Exercise B.3** Run the script puncture.py using different values for the number of grid points n_{grid}, and keep track of the time that it takes the script to run. At what value of n_{grid} will the script crash, presumably because your computer does not have enough memory to accommodate the matrix **A**? Your results should suggest why it is imperative to take advantage of the band-diagonal structure of **A** in order to produce high-resolution data.

> **Exercise B.4** The script puncture.py computes the extrinsic curvature from (3.43), and hence implements the puncture method for black holes with linear momentum. Generalize the method by considering sums of (3.41) and (3.43), thereby allowing for single black holes with both linear and angular momentum.

> **Exercise B.5** Leading-order analytical solutions to (B.37), to linear order in P^2 or J^2, are given, respectively, in Gleiser et al. (2002) and Gleiser et al. (1998).[6] Compare the solution computed by puncture.py with these analytical solutions for small values of the momenta. Explore the effects of resolution and the location of the boundary.

---

[6] See also exercises 3.12 and 3.13 in Baumgarte and Shapiro (2010).

B.3 Initial-Value Problems

B.3.1 The Method of Lines

We again start with a concrete example, namely the scalar field propagating in a vacuum, Minkowski spacetime of Section 2.1. Using definition (2.3) we brought the wave equation (2.1) into the first-order form

$$\partial_t \psi = -\kappa,$$
$$\partial_t \kappa = -D^2 \psi,$$

(B.41)

where $D^2 \psi$ is the Laplace operator acting on ψ and where we have set $\rho = 0$.

One possible starting point would be to introduce a finite-differencing grid like (B.2) for both space and time dimensions, and to express the space and time derivatives in (B.41) using function values at these spacetime grid points.[7] Many of the resulting "classical" algorithms can be written and extended within the framework of the *method of lines*. The key idea in the method of lines is to start with finite-differencing in space only, but to allow the function values at the discrete grid points x_i to remain functions of time. In one spatial dimension, a second-order finite-difference version of equations (B.41), for example, is

$$\dot{\psi}_i = -\kappa_i,$$
$$\dot{\kappa}_i = -(\psi_{i+1} - 2\psi_i + \psi_{i-1})/\Delta^2.$$

(B.42)

Here the $\kappa_i = \kappa_i(t)$ and $\psi_i = \psi_i(t)$ are considered to be functions of time, and the dot denotes a derivative with respect to time. Evidently, finite-differencing the spatial derivatives in (B.41) has turned the partial differential equations into a set of coupled ordinary differential equations (ODEs) for the function values κ_i and ψ_i. We now consider standard methods for solving these ODEs.

We will write our set of ODEs as the prototype equation

$$\dot{f} = k(t, f),$$

(B.43)

where f represents a solution vector containing all dependent functions (e.g. the κ_i and ψ_i in B.42), and k represents their time derivatives (e.g. the right-hand sides in B.42). We now want to integrate equation (B.43) from a time t_0 when $f = f_0$ to a time $t_0 + \Delta t$.

---

[7] See, e.g., the discussions in Press et al. (2007); Baumgarte and Shapiro (2010).

Our first attempt might be to use forward differencing (compare B.4) to express the time derivative in equation (B.43). This results in the *Euler method*, which we may write in the form

$$k_1 = k(t_0, f_0),$$
$$f(t + \Delta t) = f_0 + k_1 \Delta t. \tag{B.44}$$

Evaluating this for equations (B.42) yields

$$\psi_i(t_0 + \Delta t) = \psi_i(t) - \Delta t \, \kappa_i(t_0),$$
$$\kappa_i(t_0 + \Delta t) = \kappa_i(t) - (\Delta t / \Delta^2)(\psi_{i+1}(t_0) - 2\psi_i(t_0) + \psi_{i-1}(t_0)), \tag{B.45}$$

or, identifying time t_0 with a time level n and time $t_0 + \Delta t$ with $n + 1$,

$$\psi_i^{n+1} = \psi_i^n - \Delta t \, \kappa_i^n,$$
$$\kappa_i^{n+1} = \kappa_i^n - (\Delta t / \Delta^2)(\psi_{i+1}^n - 2\psi_i^n + \psi_{i-1}^n). \tag{B.46}$$

This, it turns out, is the famous *forward-time-centered-space* (FTCS) method – famous for being *unconditionally unstable* for wave equations and therefore entirely useless.[8]

We can construct a *conditionally stable* scheme by replacing (B.44) with

$$k_1 = k(t_0, f_0),$$
$$k_2 = k(t_0 + \Delta t, f_0 + k_1 \Delta t),$$
$$k_3 = k(t_0 + \Delta t, f_0 + k_2 \Delta t),$$
$$f(t + \Delta t) = f_0 + (k_1 + k_3)\Delta t / 2, \tag{B.47}$$

which results in a second-order algorithm that is equivalent to the *iterative Crank–Nicholson* scheme.[9] An alternative is the fourth-order *Runge–Kutta* scheme[10]

$$k_1 = k(t_0, f_0),$$
$$k_2 = k(t_0 + \Delta t/2, f_0 + k_1 \Delta t/2),$$
$$k_3 = k(t_0 + \Delta t/2, f_0 + k_2 \Delta t/2),$$
$$k_4 = k(t_0 + \Delta t, f_0 + k_3 \Delta t),$$
$$f(t + \Delta t) = f_0 + (k_1/6 + k_2/3 + k_3/3 + k_4/6)\Delta t / 2. \tag{B.48}$$

---

[8] The stability of time-evolution schemes can be explored with a *von Neumann* stability analysis; see, e.g., Press et al. (2007); Section 6.2.3 in Baumgarte and Shapiro (2010).
[9] The iterative step to compute k_3 is needed for stability; see Teukolsky (2000).
[10] See, e.g., equation (17.1.3) in Press et al. (2007).

Both methods are stable provided that the time step is limited by the *Courant–Friedrich–Lewy condition*

$$\Delta t < \frac{C}{v_c} \Delta. \qquad (\text{B.49})$$

Here v_c is the fastest characteristic speed (in our case the speed of light $c = 1$), the *Courant factor C* is a constant of order unity, and Δ is the spatial grid spacing.

Applying (B.47) to the second-order finite-difference equations (B.42) results in a scheme that is second order in both space and time. In order to construct methods that are higher order in time we can replace the finite differences used in (B.42) with higher-order expressions (see exercise B.2), and (B.47) with a higher-order Runge–Kutta method, e.g. (B.48). Python, for instance, comes with a large library of Runge–Kutta integrators with adaptive time step size, as do many software packages. We could use any of these for our purposes, but in many cases it may be better to write our own code.

In three spatial dimensions, spherically symmetric solutions to the scalar wave equation are given by

$$\psi = \frac{f(r \pm t)}{r}, \qquad (\text{B.50})$$

where $f(x)$ is an arbitrary function of the argument $x = r \pm t$. Here the plus sign describes ingoing waves, while the minus sign describes outgoing waves. Recognizing that, at large r, the solution is often dominated by the spherical term, we can construct an approximate *outgoing wave boundary condition* by adopting the negative sign in (B.50). This condition can be implemented in different ways. We could use (B.50) directly and compute the solution $\psi(r, t + \Delta t)$ on the boundary from an interpolation to the location $r - \Delta t$ at the previous time step, i.e.

$$\psi(r, t + \Delta t) = \frac{r - \Delta t}{r} \psi(r - \Delta t, t). \qquad (\text{B.51})$$

Alternatively we can observe that $\partial_t f = -\partial_r f$, which results in

$$\partial_t \psi = -\frac{\psi}{r} - \partial_r \psi = -\frac{\psi}{r} - l^i \partial_i \psi, \qquad (\text{B.52})$$

with $l^i = x^i / r$, where the last term, $l^i \partial_i \psi$, evaluates the radial derivative $\partial_r \psi$ in Cartesian coordinates. While this condition is exact only for spherical scalar waves originating from the origin, it is often adopted as an approximate outgoing-wave boundary condition for other waves at large r as well.

B.3.2 A Worked Example: Maxwell's Equations

As in Section B.2.3 we will provide a concrete example to demonstrate how all the above can be put together. Specifically, consider Maxwell's equations, both in the original form (4.1), (4.2) and the reformulated version (4.8)–(4.10) (with $\eta = 1$), in vacuum (i.e. $\rho = 0$ and $j^i = 0$). We will integrate these equations in three spatial dimensions using Cartesian coordinates, imposing the outgoing wave boundary condition (B.52) on the outer boundaries at $x_{\max}$, $y_{\max}$, and $z_{\max}$.

Choosing the gauge variable to vanish, i.e. $\phi = 0$, an analytical dipolar solution to Maxwell's equations in vacuum is given in spherical polar coordinates by

$$A^{\hat\varphi} = \mathcal{A}\sin\theta\left(\frac{e^{-(v/\lambda)^2} - e^{-(u/\lambda)^2}}{(r/\lambda)^2} - 2\frac{ve^{-(v/\lambda)^2} + ue^{-(u/\lambda)^2}}{r}\right),$$

$$A^{\hat r} = A^{\hat\theta} = 0, \tag{B.53}$$

where $\mathcal{A}$ parameterizes the amplitude of the wave, λ is a constant with units of length, and u and v are combinations given by $u = t + r$ and $v = t - r$. From the orthonormal toroidal component $A^{\hat\varphi}$ we can find the Cartesian components of the fields using the transformation

$$A^x = -A^{\hat\varphi}\sin\varphi, \quad A^y = A^{\hat\varphi}\cos\varphi, \quad A^z = 0. \tag{B.54}$$

As initial data we will choose the solution (B.53) at $t = 0$, when $A^i = 0$ and

$$E^{\hat\varphi} = -\partial_t A^{\hat\varphi} = -8\mathcal{A}\frac{r\sin\theta}{\lambda^2}\exp(-(r/\lambda)^2) \qquad (t = 0). \tag{B.55}$$

We then evolve these data subject to the familiar Lorenz gauge condition. This gives

$$\partial_t\phi = -D_i A^i = -\Gamma \tag{B.56}$$

(see also equation 4.6), which is consistent with $\phi = 0$ for the solution (B.53). Note also that in Cartesian coordinates in flat spacetimes we do not need to distinguish between contravariant and covariant components, i.e. $A_i = A^i$.

Table B.1 Symmetries of the dynamical
variables across the $x = 0$, $y = 0$, and
$z = 0$ planes.

	$x = 0$	$y = 0$	$z = 0$
E^x, A^x	sym	anti	sym
E^y, A^y	anti	sym	sym
E^z, A^z	sym	sym	anti
ϕ, Γ	anti	anti	sym

In order to reduce the size of our numerical grid we will take advantage of the symmetry of the solution (B.53). Specifically, we will evolve the solution in only one octant (namely the octant in which x, y, and z are all positive) and will impose symmetry boundary conditions on the coordinate planes $x = 0$, $y = 0$, and $z = 0$. For the x- and y- components of the fields, these conditions can be found by inspection from (B.53) and (B.54), and for the z-components we impose axisymmetry together with equatorial symmetry across the $z = 0$ plane. We summarize the resulting conditions in Table B.1. For example, the boundary conditions across $z = 0$ imply $E^x|_{z=0^+} = E^x|_{z=0^-}$, $E^z|_{z=0^+} = -E^z|_{z=0^-}$, and so forth.

When we integrate Maxwell's equations in the original formulation we encounter in equation (4.2) the gradient of the divergence, $D_i D_j A^j$, or, in Cartesian coordinates, $\partial_i \partial_j A^j$. It is tempting to compute first the divergence $\partial_j A^j$ and then the gradient of the result. That would be a bad idea, since it would extend the finite-differencing stencil and hence would require more neighboring grid points. It is much better to express this operator in terms of second derivatives. For $i = x$, for example, we have

$$\partial_x \partial_j A^j = \partial_x \partial_x A^x + \partial_x \partial_y A^y + \partial_x \partial_z A^z. \tag{B.57}$$

Finite-differencing this to second order, we can use (B.13) for the second derivative in the first term on the right-hand side. The last two terms on the right-hand side, however, involve mixed derivatives, which we have not yet encountered. We can finite-difference these terms by using (B.12) twice. Assuming a uniform grid with equal grid spacing Δ in all dimensions, as in Section B.2.1, we obtain for the middle term on the right-hand side

$$(\partial_x \partial_y A^y)_{ijk} = \frac{1}{4\Delta^2}\Big(A^y_{i+1,j+1,k} - A^y_{i-1,j+1,k} - A^y_{i+1,j-1,k} + A^y_{i-1,j-1,k} \Big),$$
$$(B.58)$$

and a similar expression for the last term in (B.57).

During the evolution we monitor the constraint violation

$$\mathcal{C} = \partial_i E^i \tag{B.59}$$

and compute its L2-norm

$$||\mathcal{C}|| = \left(\int \mathcal{C}^2 d^3x \right)^{1/2} \tag{B.60}$$

as a test of our code – more on that in Section B.4 below.

As in Section B.2.3 we can again eliminate some constants from the problem. We first observe that Maxwell's equations are linear, so that the amplitude $\mathcal{A}$ completely drops out of the problem if we rescale the fields according to $\tilde{A}^i = A^i/\mathcal{A}$ and $\tilde{E}^i = E^i/\mathcal{A}$ as well as $\tilde{\Gamma} = \Gamma/\mathcal{A}$ and $\tilde{\phi} = \phi/\mathcal{A}$. Similarly, the physical unit of length enters the problem only through the constant λ. We may therefore rescale lengths and times (recall that $c = 1$) with λ; specifically, we introduce dimensionless coordinates $\bar{x}^i = x^i/\lambda$ and $\bar{t} = t/\lambda$ as well as dimensionless fields $\bar{E}^i = \lambda \tilde{E}^i$ and $\bar{\Gamma} = \lambda \tilde{\Gamma}$. Since $\tilde{A}^i$ and $\tilde{\phi}$ are dimensionless already, we will set $\bar{A}^i = \tilde{A}^i$ and $\bar{\phi} = \tilde{\phi}$. So, writing our code in terms of the quantities with overbars, the constants $\mathcal{A}$ and λ will not make an appearance – of course, we could achieve the same effect by simply setting $\mathcal{A}$ and λ to unity. To summarize, we use the variables

$$\begin{aligned}
\bar{t} &= t/\lambda, & \bar{x}^i &= x^i/\lambda, & \bar{A}^i &= A^i/\mathcal{A}, \\
\bar{E}^i &= E^i\lambda/\mathcal{A}, & \bar{\phi} &= \phi/\mathcal{A}, & \bar{\Gamma} &= \Gamma\lambda/\mathcal{A},
\end{aligned} \tag{B.61}$$

and recognize that we can always recover the physical variables for arbitrary values of $\mathcal{A}$ and λ from the above scalings.

Exercise B.6 Apply the above arguments to find the corresponding scalings

$$\bar{\mathcal{C}} = \mathcal{C}\lambda^2/\mathcal{A}, \qquad ||\bar{\mathcal{C}}|| = ||\mathcal{C}||\lambda^{1/2}/\mathcal{A} \tag{B.62}$$

for the constraint violation $\mathcal{C}$ and its norm.

We list below a Python script `maxwell.py` that implements the above method, but it can also be downloaded from www.cambridge. org/NRStartingfromScratch. Like the script `puncture.py` in Section B.2.3, `maxwell.py` employs a hierarchy of class structures.

The class `OperatorsOrder2` contains all the derivative operators that we will need. A base class `Maxwell`, which instantiates `OperatorsOrder2`, contains all the methods that can be used for the integration of both the original Maxwell equations (4.1) and (4.2) and the reformulated Maxwell equations (4.8)–(4.10). Two derived classes, `Original` and `Reformulated`, contain methods that are specific to either the original or the reformulated Maxwell equations. Either of these two classes can be instantiated in `main` for a number of different input parameters.

```python
"""Routines that are used to solve Maxwell's equations."""

import sys
from numpy import zeros, linspace, exp, sqrt

class OperatorsOrder2:
    """ This class contains all of the derivative operators that we will
    ever need. The operators are implemented to the second order.
    """

    def __init__(self, n_grid, delta):
        """Constructor for class OperatorsOrder2."""

        print(" Setting up 2nd-order derivative operators ")
        self.n_grid = n_grid
        self.delta = delta

    def laplace(self, fct):
        """Compute Laplace operator of function fct in interior of grid."""

        n_grid = self.n_grid
        ood2 = 1.0 / self.delta ** 2
        lap = zeros((n_grid, n_grid, n_grid))

        for i in range(1, n_grid - 1):
            for j in range(1, n_grid - 1):
                for k in range(1, n_grid - 1):
                    # see (B.26)
                    dfddx = (fct[i + 1, j, k] - 2.0 * fct[i, j, k] +
                             fct[i - 1, j, k])
                    dfddy = (fct[i, j + 1, k] - 2.0 * fct[i, j, k] +
                             fct[i, j - 1, k])
                    dfddz = (fct[i, j, k + 1] - 2.0 * fct[i, j, k] +
                             fct[i, j, k - 1])
                    lap[i, j, k] = ood2 * (dfddx + dfddy + dfddz)

        return lap

    def gradient(self, fct):
        """Compute gradient of function fct in interior of grid."""

        n_grid = self.n_grid
        oo2d = 0.5 / self.delta
        grad_x = zeros((n_grid, n_grid, n_grid))
        grad_y = zeros((n_grid, n_grid, n_grid))
        grad_z = zeros((n_grid, n_grid, n_grid))

        for i in range(1, n_grid - 1):
            for j in range(1, n_grid - 1):
                for k in range(1, n_grid - 1):
                    # see (A.12)
```

```
                grad_x[i, j, k] = oo2d * (fct[i + 1, j, k] -
                                          fct[i - 1, j, k])
                grad_y[i, j, k] = oo2d * (fct[i, j + 1, k] -
                                          fct[i, j - 1, k])
                grad_z[i, j, k] = oo2d * (fct[i, j, k + 1] -
                                          fct[i, j, k - 1])

        return grad_x, grad_y, grad_z


    def divergence(self, v_x, v_y, v_z):
        """Compute divergence of vector v in interior of grid."""

        n_grid = self.n_grid
        oo2d = 0.5 / self.delta
        div = zeros((n_grid, n_grid, n_grid))

        for i in range(1, n_grid - 1):
            for j in range(1, n_grid - 1):
                for k in range(1, n_grid - 1):
                    # see (B.12)
                    v_xx = v_x[i + 1, j, k] - v_x[i - 1, j, k]
                    v_yy = v_y[i, j + 1, k] - v_y[i, j - 1, k]
                    v_zz = v_z[i, j, k + 1] - v_z[i, j, k - 1]
                    div[i, j, k] = oo2d * (v_xx + v_yy + v_zz)

        return div


    def grad_div(self, A_x, A_y, A_z):
        """Compute gradient of divergence in interior of grid."""

        n_grid = self.n_grid
        ood2 = 1.0 / self.delta ** 2
        grad_div_x = zeros((n_grid, n_grid, n_grid))
        grad_div_y = zeros((n_grid, n_grid, n_grid))
        grad_div_z = zeros((n_grid, n_grid, n_grid))

        for i in range(1, n_grid - 1):
            for j in range(1, n_grid - 1):
                for k in range(1, n_grid - 1):
                    # xx_A_x denotes \partial_x \partial_x A_x, etc...

                    xx_A_x = (A_x[i + 1, j, k] - 2.0 * A_x[i, j, k] +
                              A_x[i - 1, j, k])
                    xy_A_y = (A_y[i + 1, j + 1, k] - A_y[i + 1, j - 1, k] -
                              A_y[i - 1, j + 1, k] + A_y[i - 1, j - 1, k])
                    xz_A_z = (A_z[i + 1, j, k + 1] - A_z[i + 1, j, k - 1] -
                              A_z[i - 1, j, k + 1] + A_z[i - 1, j, k - 1])
                    grad_div_x[i, j, k] = ood2 * (xx_A_x + 0.25 * xy_A_y +
                                                  0.25 * xz_A_z)
                    yx_A_x = (A_x[i + 1, j + 1, k] - A_x[i + 1, j - 1, k] -
                              A_x[i - 1, j + 1, k] + A_x[i - 1, j - 1, k])
                    yy_A_y = (A_y[i, j + 1, k] - 2.0 * A_y[i, j, k] +
                              A_y[i, j - 1, k])
                    yz_A_z = (A_z[i, j + 1, k + 1] - A_z[i, j + 1, k - 1] -
                              A_z[i, j - 1, k + 1] + A_z[i, j - 1, k - 1])
                    grad_div_y[i, j, k] = ood2 * (0.25 * yx_A_x + yy_A_y +
                                                  0.25 * yz_A_z)
                    zx_A_x = (A_x[i + 1, j, k + 1] - A_x[i + 1, j, k - 1] -
                              A_x[i - 1, j, k + 1] + A_x[i - 1, j, k - 1])
                    zy_A_y = (A_y[i, j + 1, k + 1] - A_y[i, j + 1, k - 1] -
                              A_y[i, j - 1, k + 1] + A_y[i, j - 1, k - 1])
                    zz_A_z = (A_z[i, j, k + 1] - 2.0 * A_z[i, j, k] +
                              A_z[i, j, k - 1])
                    grad_div_z[i, j, k] = ood2 * (0.25 * zx_A_x + 0.25 *
                                                  zy_A_y + zz_A_z)
```

```
        return grad_div_x, grad_div_y, grad_div_z

    def partial_derivs(self, fct, i, j, k):
        """Compute partial derivatives. Unlike all
        other operators in this class, this function returns derivatives
        at one grid point (i,j,k) only, but including 'points on the
                                                    boundaries."""

        oo2d = 0.5 / self.delta

        # use (A.10) in interior, and (A.5) on boundaries
        # partial f / partial x
        if i == 0:
            f_x = oo2d * (-3.0 * fct[i, j, k] + 4.0 * fct[i + 1, j, k] -
                          fct[i + 2, j, k])
        elif i == self.n_grid - 1:
            f_x = oo2d * (3.0 * fct[i, j, k] - 4.0 * fct[i - 1, j, k] +
                          fct[i - 2, j, k])
        else:
            f_x = oo2d * (fct[i + 1, j, k] - fct[i - 1, j, k])

        # partial f / partial y
        if j == 0:
            f_y = oo2d * (-3.0 * fct[i, j, k] + 4.0 * fct[i, j + 1, k] -
                          fct[i, j + 2, k])
        elif j == self.n_grid - 1:
            f_y = oo2d * (3.0 * fct[i, j, k] - 4.0 * fct[i, j - 1, k] +
                          fct[i, j - 2, k])
        else:
            f_y = oo2d * (fct[i, j + 1, k] - fct[i, j - 1, k])

        # partial f / partial z
        if k == 0:
            f_z = oo2d * (-3.0 * fct[i, j, k] + 4.0 * fct[i, j, k + 1] -
                          fct[i, j, k + 2])
        elif k == self.n_grid - 1:
            f_z = oo2d * (3.0 * fct[i, j, k] - 4.0 * fct[i, j, k - 1] +
                          fct[i, j, k - 2])
        else:
            f_z = oo2d * (fct[i, j, k + 1] - fct[i, j, k - 1])
        return f_x, f_y, f_z


class Maxwell:
    """ The base class for evolution of Maxwell's equations.
    """

    def __init__(self, n_grid, x_out, filename_stem, n_vars):
        """Constructor sets up coordinates, memory for variables, and
        opens output file.
        """

        print(" Initializing class Maxwell with n_grid = ", n_grid,
              ", x_out = ", x_out)
        self.n_grid = n_grid
        self.filename_stem = filename_stem
        self.n_vars = n_vars
        self.delta = float(x_out) / (n_grid - 2.0)
        delta = self.delta
        print("     grid spacing delta =",delta)

        # set up cell-centered grid on interval (0, x_out) in each
        # dimension, but pad grid with one ghost zone.  Will use symmetry
        # on "inner" boundaries, and out-going wave conditions on outer
        # boundaries.
        self.x = linspace(-delta / 2.0, x_out + delta / 2.0, n_grid)
        self.y = linspace(-delta / 2.0, x_out + delta / 2.0, n_grid)
```

```python
        self.z = linspace(-delta / 2.0, x_out + delta / 2.0, n_grid)
        self.r = zeros((n_grid, n_grid, n_grid))

        for i in range(0, n_grid):
            for j in range(0, n_grid):
                for k in range(0, n_grid):
                    self.r[i, j, k] = sqrt(
                        self.x[i] ** 2 + self.y[j] ** 2 + self.z[k] ** 2
                    )

        # set up derivative operators
        self.ops = OperatorsOrder2(n_grid, delta)

        # set up all variables common to both approaches
        self.E_x = zeros((n_grid, n_grid, n_grid))
        self.E_y = zeros((n_grid, n_grid, n_grid))
        self.E_z = zeros((n_grid, n_grid, n_grid))
        self.A_x = zeros((n_grid, n_grid, n_grid))
        self.A_y = zeros((n_grid, n_grid, n_grid))
        self.A_z = zeros((n_grid, n_grid, n_grid))
        self.phi = zeros((n_grid, n_grid, n_grid))
        self.constraint = zeros((n_grid, n_grid, n_grid))

        # open file that records constraint violations
        filename = self.filename_stem + "_constraints.data"
        self.constraint_file = open(filename, "w")
        if self.constraint_file:
            self.constraint_file.write(
                "# Constraint violations and errors with n = %d\n" % (n_grid)
            )
            self.constraint_file.write("# t         C_norm\n")
            self.constraint_file.write("#========================\n")
        else:
            print(" Could not open file",filename," for output")
            print(" Check permissions?")
            sys.exit(2)
        # keep track of time
        self.t = 0.0


    def __del__(self):
        """Close output file in destructor."""
        if self.constraint_file:
            self.constraint_file.close()
    def initialize(self):
        """Set up initial data for E^i (assuming A^i = 0 initially);
        see (B.55).
        """

        for i in range(0, self.n_grid):
            for j in range(0, self.n_grid):
                for k in range(0, self.n_grid):
                    rl = self.r[i, j, k]
                    costheta = self.z[k] / rl
                    sintheta = sqrt(1.0 - costheta ** 2)
                    rho = sqrt(self.x[i] ** 2 + self.y[j] ** 2)
                    cosphi = self.x[i] / rho
                    sinphi = self.y[j] / rho
                    E_phi = -8.0 * rl * sintheta * exp(- rl ** 2)
                    self.E_x[i, j, k] = -E_phi * sinphi
                    self.E_y[i, j, k] = E_phi * cosphi
                    self.E_z[i, j, k] = 0.0

        self.check_constraint()


    def check_constraint(self):
        """Check constraint violation."""
```

```
        self.constraint = self.ops.divergence(self.E_x, self.E_y, self.E_z)
        norm_c = 0.0

        for i in range(1, self.n_grid - 1):
            for j in range(1, self.n_grid - 1):
                for k in range(1, self.n_grid - 1):
                    norm_c += self.constraint[i, j, k] ** 2

        norm_c = sqrt(norm_c * self.delta ** 3)
        self.constraint_file.write("%e %e \n" % (self.t, norm_c))
        print(" Constraint violation at time {0:.3f} : {1:.4e}".
                 format(self.t, norm_c))
        return norm_c


def integrate(self, courant, t_max, t_check):
    """Carry out time integration."""

    print(" Integrating to time",t_max,"with Courant factor",courant)
    print("     checking results after times", t_check)
    self.write_data()
    while self.t + t_check <= t_max:
        fields = self.wrap_up_fields()
        # call stepper to integrate from current time t to t + t_const
        self.t, fields = self.stepper(fields, courant, t_check)
        self.unwrap_fields(fields)
        self.check_constraint()
        self.write_data()


def stepper(self, fields, courant, t_const):
    """Stepper for Runge-Kutta integrator, integrates
    from current time t to t + t_const by repeatedly calling icn.
    """
    delta_t = courant * self.delta
    time = self.t
    t_fin = time + t_const

    while time < t_fin:
        if t_fin - time < delta_t:
            delta_t = t_fin - time
        # call icn to carry out one time step

        fields = self.icn(fields, delta_t)
        time += delta_t

    return time, fields


def icn(self, fields, dt):
    """Carry out one 2nd-order iterative Crank-Nicholson step;
    see (B.47).
    The routine derivatives(fields) is defined in the derived classes
    Original and Reformulated, and provides time derivatives of the fields
    according to the two different versions of Maxwell's equations.
    The routine update_fields(fields, fields_dot, factor, dt) adds
    factor * dt * fields_dot to fields.
    """

    # Step 1: get derivs at t0 and update fields with half-step
    fields_dot = self.derivatives(fields)
    new_fields = self.update_fields(fields, fields_dot, 0.5, dt)

    # Step 2: get derivs at t = t0 + dt and update fields temporarily
    fields_temp = self.update_fields(fields, fields_dot, 1.0, dt)
    fields_dot = self.derivatives(fields_temp)
```

```
        # Step 3: do it again...  (iterative step)
        fields_temp = self.update_fields(fields, fields_dot, 1.0, dt)
        fields_dot = self.derivatives(fields_temp)

        # Finally: update new_fields with second half-step
        new_fields = self.update_fields(new_fields, fields_dot, 0.5, dt)
        #

        return new_fields


    def update_fields(self, fields, fields_dot, factor, dt):
        """Updates all fields in the list fields,
        by adding factor * fields_dot * dt.
        """

        new_fields = []
        for var in range(0, self.n_vars):
            field = fields[var]
            field_dot = fields_dot[var]
            new_field = field + factor * field_dot * dt
            new_fields.append(new_field)

        return new_fields

    def outgoing_wave(self, f_dot, f):
        """Computes time derivatives of fields from
        outgoing-wave boundary condition;
        see  (B.52)
        """
        n_grid = self.n_grid

        i = n_grid - 1  # upper x-boundary
        for j in range(0, n_grid - 1):
            for k in range(0, n_grid - 1):
                f_x, f_y, f_z = self.ops.partial_derivs(f, i, j, k)
                f_dot[i, j, k] = (-(f[i, j, k] + self.x[i] * f_x +
                                   self.y[j] * f_y + self.z[k] * f_z)
                                  / self.r[i, j, k])

        j = n_grid - 1  # upper y-boundary
        for i in range(0, n_grid):
            for k in range(0, n_grid - 1):

                f_x, f_y, f_z = self.ops.partial_derivs(f, i, j, k)
                f_dot[i, j, k] = (-(f[i, j, k] + self.x[i] * f_x +
                                   self.y[j] * f_y + self.z[k] * f_z)
                                  / self.r[i, j, k])

        k = n_grid - 1  # upper z-boundary
        for i in range(0, n_grid):
            for j in range(0, n_grid):
                f_x, f_y, f_z = self.ops.partial_derivs(f, i, j, k)
                f_dot[i, j, k] = (-(f[i, j, k] + self.x[i] * f_x +
                                   self.y[j] *  f_y + self.z[k] * f_z)
                                  / self.r[i, j, k])

    def symmetry(self, f_dot, x_sym, y_sym, z_sym):
        """Computes time derivatives on inner boundaries from symmetry,"""

        n_grid = self.n_grid

        i = 0  # lower x-boundary
        for j in range(1, n_grid - 1):
            for k in range(1, n_grid - 1):
                f_dot[i, j, k] = x_sym * f_dot[i + 1, j, k]
```

```
        j = 0  # lower y-boundary
        for i in range(0, n_grid - 1):
            for k in range(1, n_grid - 1):
                f_dot[i, j, k] = y_sym * f_dot[i, j + 1, k]

        k = 0  # lower z-boundary
        for i in range(0, n_grid - 1):
            for j in range(0, n_grid - 1):
                f_dot[i, j, k] = z_sym * f_dot[i, j, k + 1]


    def write_data(self):
        """Write data to file."""

        HEADER1 = "# x            y            E_x          E_y          "
        HEADER2 = "A_x          A_y          phi          C\n"
        WRITER = "%e  %e  %e  %e  %e  %e  %e  %e\n"

        filename = self.filename_stem + "_fields_{0:.3f}.data".format(self.t)
        out = open(filename, "w")
        if out:
            k = 1
            out.write("# data at time {0:.3f} at z = {1:e} \n".
                      format(self.t, self.z[k]))
            out.write(HEADER1)
            out.write(HEADER2)
            n_grid = self.n_grid

            for i in range(1, n_grid):
                for j in range(1, n_grid):
                    out.write(WRITER % (self.x[i], self.y[j],
                                        self.E_x[i, j, k], self.E_y[i, j, k],
                                        self.A_x[i, j, k], self.A_y[i, j, k],
                                        self.phi[i, j, k],
                                        self.constraint[i, j, k],))
                out.write("\n")
            out.close()
        else:
            print(" Could not open file", filename,"in write_data()")
            print(" Check permissions?")
        return


class Original(Maxwell):
    """ Derived class Original contains methods
    for the original version of Maxwell's equations.

    Note: in this version, we integrate n_vars = 7 independent variables.
    """

    def __init__(self, n_grid, x_out):
        """Constructor doesn't do much..."""

        print(" Integrating original Maxwell Equations ")
        print("    using n_grid =", n_grid, "\b^3 grid points with ")
        print("    outer boundary at x_out =", x_out)
        IND_VARS = 7
        filename_stem = "Max_orig_" + str(n_grid) + "_" + str(x_out)
        Maxwell.__init__(self, n_grid, x_out, filename_stem, IND_VARS)


    def wrap_up_fields(self):
        """Bundles up the seven variables into the list fields."""

        return (self.E_x, self.E_y, self.E_z, self.A_x, self.A_y,
                self.A_z, self.phi)
```

```python
def unwrap_fields(self, fields):
    """Extracts the independent variables from the list fields."""

    self.E_x = fields[0]
    self.E_y = fields[1]
    self.E_z = fields[2]
    self.A_x = fields[3]
    self.A_y = fields[4]
    self.A_z = fields[5]
    self.phi = fields[6]


def derivatives(self, fields):
    """Computes the time derivatives of the fields: Maxwell's equations."""

    self.unwrap_fields(fields)

    #
    # compute derivatives
    #
    Lap_A_x = self.ops.laplace(self.A_x)
    Lap_A_y = self.ops.laplace(self.A_y)
    Lap_A_z = self.ops.laplace(self.A_z)
    grad_x_phi, grad_y_phi, grad_z_phi = self.ops.gradient(self.phi)
    grad_div_x_A, grad_div_y_A, grad_div_z_A = self.ops.grad_div(
        self.A_x, self.A_y, self.A_z)
    DivA = self.ops.divergence(self.A_x, self.A_y, self.A_z)

    #
    # then compute time derivatives from original version
    # of Maxwell's equations,
    # see (4.1) and (4.2)
    #
    A_x_dot = -self.E_x - grad_x_phi
    A_y_dot = -self.E_y - grad_y_phi
    A_z_dot = -self.E_z - grad_z_phi
    E_x_dot = -Lap_A_x + grad_div_x_A
    E_y_dot = -Lap_A_y + grad_div_y_A
    E_z_dot = -Lap_A_z + grad_div_z_A
    phi_dot = -DivA


    #
    # finally fix inner boundaries from symmetry, see Table B.1...
    #
    self.symmetry(E_x_dot, 1, -1, 1)
    self.symmetry(E_y_dot, -1, 1, 1)
    self.symmetry(E_z_dot, 1, 1, -1)
    self.symmetry(A_x_dot, 1, -1, 1)
    self.symmetry(A_y_dot, -1, 1, 1)
    self.symmetry(A_z_dot, 1, 1, -1)
    self.symmetry(phi_dot, -1, -1, 1)

    #
    # ...and outer boundaries using outgoing-wave boundary condition (B.52)
    #
    self.outgoing_wave(E_x_dot, self.E_x)
    self.outgoing_wave(E_y_dot, self.E_y)
    self.outgoing_wave(E_z_dot, self.E_z)
    self.outgoing_wave(A_x_dot, self.A_x)
    self.outgoing_wave(A_y_dot, self.A_y)
    self.outgoing_wave(A_z_dot, self.A_z)
    self.outgoing_wave(phi_dot, self.phi)

    #
    return (E_x_dot, E_y_dot, E_z_dot, A_x_dot, A_y_dot, A_z_dot, phi_dot)
```

```python
class Reformulated(Maxwell):
    """ Derived class Reformulated contains methods for
    the reformulated version of Maxwell's equations.

    Note: in the reformulated version we integrate n_vars = 8
    independent variables
    """

    def __init__(self, n_grid, x_out):
        """Constructor doesn't do much..."""

        IND_VARS = 8
        print(" Integrating reformulated Maxwell Equations ")
        print("    using n_grid =", n_grid, "\b^3 grid points with ")
        print("    outer boundary at x_out =", x_out)

        filename_stem = "Max_reform_" + str(n_grid) + "_" + str(x_out)
        Maxwell.__init__(self, n_grid, x_out, filename_stem, IND_VARS)

        # additional independent variable Gamma:
        self.Gamma = zeros((n_grid, n_grid, n_grid))


    def wrap_up_fields(self):
        """Bundles up the eight variables into the list fields."""

        return (self.E_x, self.E_y, self.E_z, self.A_x, self.A_y, self.A_z,
                self.phi, self.Gamma,)


    def unwrap_fields(self, fields):
        """Extracts the independent variables from the list fields."""

        self.E_x = fields[0]
        self.E_y = fields[1]
        self.E_z = fields[2]
        self.A_x = fields[3]
        self.A_y = fields[4]
        self.A_z = fields[5]
        self.phi = fields[6]
        self.Gamma = fields[7]

    def derivatives(self, fields):
        """Computes the time derivatives of the fields: Maxwell's equations."""

        self.unwrap_fields(fields)

        #
        # compute derivatives
        #
        Lap_A_x = self.ops.laplace(self.A_x)
        Lap_A_y = self.ops.laplace(self.A_y)
        Lap_A_z = self.ops.laplace(self.A_z)
        grad_x_Gamma, grad_y_Gamma, grad_z_Gamma = \
          self.ops.gradient(self.Gamma)
        grad_x_phi, grad_y_phi, grad_z_phi = self.ops.gradient(self.phi)
        Lap_phi = self.ops.laplace(self.phi)

        #
        # then compute time derivatives from reformulated version
        # of Maxwell's equations;
        # see equations (4.8) through (4.10)
        #
        A_x_dot = -self.E_x - grad_x_phi
        A_y_dot = -self.E_y - grad_y_phi
        A_z_dot = -self.E_z - grad_z_phi
```

```
            E_x_dot = -Lap_A_x + grad_x_Gamma
            E_y_dot = -Lap_A_y + grad_y_Gamma
            E_z_dot = -Lap_A_z + grad_z_Gamma
            phi_dot = -self.Gamma
            Gamma_dot = -Lap_phi

            #
            # finally fix inner boundaries from symmetry, see Table B.1
            #
            self.symmetry(E_x_dot, 1, -1, 1)
            self.symmetry(E_y_dot, -1, 1, 1)
            self.symmetry(E_z_dot, 1, 1, -1)
            self.symmetry(A_x_dot, 1, -1, 1)
            self.symmetry(A_y_dot, -1, 1, 1)
            self.symmetry(A_z_dot, 1, 1, -1)
            self.symmetry(phi_dot, -1, -1, 1)
            self.symmetry(Gamma_dot, -1, -1, 1)

            #
            # ...and outer boundaries from outgoing-wave boundary condition (B.52)
            #
            self.outgoing_wave(E_x_dot, self.E_x)
            self.outgoing_wave(E_y_dot, self.E_y)
            self.outgoing_wave(E_z_dot, self.E_z)
            self.outgoing_wave(A_x_dot, self.A_x)
            self.outgoing_wave(A_y_dot, self.A_y)
            self.outgoing_wave(A_z_dot, self.A_z)
            self.outgoing_wave(phi_dot, self.phi)
            self.outgoing_wave(Gamma_dot, self.Gamma)

            #
            return (E_x_dot, E_y_dot, E_z_dot, A_x_dot, A_y_dot, A_z_dot, phi_dot,
                Gamma_dot,)

#
#============================================================================
# Main routine
#============================================================================
#
def main():
    """Main routine: define parameters and carry out integration."""
    print(" -------------------------------------------------")
    print(" --- maxwell.py --- use flag -h for list of options ---")
    print(" -------------------------------------------------")
    #
    # set default values for variables
    #
    # number of grid points
    n_grid = 26
    # location of outer boundary
    x_out = 6.0
    # Courant factor
    courant = 0.5
    # integrate to time t_max
    t_max = 15
    # check constraints at increments of t_const
    t_check = 0.1
    # integrate original equations
    original = True
    #
    # now look for flags to overwrite default values
    #
    for i in range(len(sys.argv)):
        if sys.argv[i] == "-h":
            usage()
            return
        if sys.argv[i] == "-n_grid":
            n_grid = int(sys.argv[i+1])
```

```
        if sys.argv[i] == "-x_out":
            x_out = float(sys.argv[i+1])
        if sys.argv[i] == "-reform":
            original = False
        if sys.argv[i] == "-courant":
            courant = float(sys.argv[i+1])
        if sys.argv[i] == "-t_max":
            t_max = float(sys.argv[i+1])
        if sys.argv[i] == "-t_check":
            t_check = float(sys.argv[i+1])
    # decide whether to integrate original or reformulated equations
    if original:
        james_clerk = Original(n_grid, x_out)
    else:
        james_clerk = Reformulated(n_grid, x_out)
    #
    # set up initial data
    james_clerk.initialize()
    #
    # now integrate...
    james_clerk.integrate(courant, t_max, t_check)


def usage():
    print("Solves Maxwell's equations in vacuum for electromagnetic wave.")
    print("")
    print("The following options can be used to overwrite default parameters")
    print("\t-n_grid: number of grid points [Default: 26 ]")
    print("\t-x_out: location of outer boundary [6.0]")
    print("\t-reform: use reformulated rather than original equations")
    print("\t-courant: Courant factor [0.5]")
    print("\t-t_max: maximum time [15]")
    print("\t-t_check: time after which constraints data are written [0.1]")
    print("For example, to evolve reformulated equations with x_out = 6.0, call")
    print("\tpython3 maxwell.py  rcform -x_out 6.0")

if __name__ == '__main__':
    main()
```

This script can be downloaded from www.cambridge.org/NRStarting fromScratch. Saving the routine as `maxwell.py`, it can be run with the command

```
python3 maxwell.py
```

to evolve the initial data (B.55). Executed as above, without any options, the script will adopt a number of default options, for example for the number of grid points `<n_grid>` (set to 26, which includes two ghost points) and the location of the outer boundary `x_out` (set to 6.0, in units of λ). By default, the script also evolves the original version of Maxwell's equations (equations 4.1 and 4.2). To see all options, their default values, and how to overwrite them, run `maxwell.py` with the flag `-h`. For example, to evolve the reformulated set of Maxwell's equations (equations 4.8–4.10) to a time `t_max = 150`, storing data at intervals of `t_check = 1.0`, run `maxwell.py` with

```
python3 maxwell.py -reform -t_max 150.0 -t_check 1.0
```

The script will produce two different types of output files. One output file, named `Max_orig_<n_grid>_<x_out>_constraints.data`, lists the L2-norm of the constraint violations $\mathcal{C}$ as a function of time (see Fig. 4.1), evaluated at time intervals `t_check`. A second set of output files, `Max_reform_<n_grid>_<x_out>_fields_<t>.data`, lists the electromagnetic fields as well as the constraint violations as a function of x and y for $z = z_1$ at times `<t>`; these files are also created at time intervals `t_check`. Examples of the former output are shown in Figs. 4.1 and B.2. The latter data files can be used to produce animations of the electromagnetic field. To do so, download the script `maxwell_animation.py`, also available at www.cambridge.org/NRStartingfromScratch, and run it with

```
python3 maxwell_animation.py
```

We again caution the reader that results may depend on the platform and installations of Python and its libraries; if Python3 does not work, earlier versions may. The script `maxwell_animation.py` can also be run with a number of different options (as before, they are listed when the flag `-h` is used).

Snapshots of the electric field $\bar{E}^x$ at some representative times are shown in Fig. B.1; they clearly show the electromagnetic wave propagating through the numerical grid and disappearing through the outer boundaries.

Exercise B.7 Verify that the script `maxwell.py` becomes unstable and produces unphysical results if the Courant factor is chosen too large.

Figure 4.1 shows for comparison the constraint violations $||\bar{\mathcal{C}}||$ in the original and reformulated Maxwell equations, confirming the expectation of Section 4.1.1 that, at late times, constraint violations will remain constant in the original formulation but will propagate off the grid and hence decrease in the reformulated version. As we discussed in Section 4.1.1, this decrease is exponential, which can be interpreted as being due to a certain fraction of the constraint violations leaving the grid in each light-crossing time.

Exercise B.8 Explore the dependence of the rate of exponential decay of constraint violations in the reformulated version on the location of the outer boundary, $\bar{x}_{out}$. In order to confirm that the decay constant is indeed proportional to the light-crossing time, run the script `maxwell.py` for different values of $\bar{x}_{out}$ and plot $||\bar{\mathcal{C}}||$ as a function of t/x_{out}. Adjust n_{grid} when changing $\bar{x}_{out}$ so that the resolution $\Delta = \bar{x}_{out}/(n_{grid} - 2)$ remains constant.

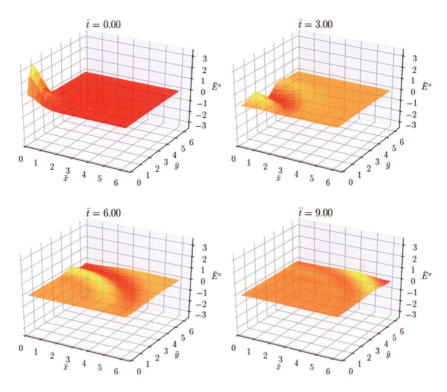

Figure B.1 The dynamical evolution of an electromagnetic wave. The panels show snapshots of the electric field $\bar{E}^x$ at selected times, starting with the initial data (B.55) for $\bar{t} = 0$. The evolution was performed with the script maxwell.py using default parameters. The images were then produced from the output data files with the script maxwell_animation.py.

B.4 Convergence Tests

After writing a numerical code it is important to verify that it produces correct results. In some cases we may be able to run the code for a case in which an analytical solution is known. In this situation we can compute the numerical error directly by computing the difference between the code's results with the analytical solution. Even in the absence of an analytical solution we may be able to monitor auxiliary constraints that the solution has to satisfy, for example $\mathcal{C} = 0$ in the case of electrodynamics in Section B.3.2. Usually, though, there will be some numerical error. We hope, then, that these errors are small in some sense, but in what sense? To make this precise, we should consider the *convergence* of numerical schemes, which we will now describe.

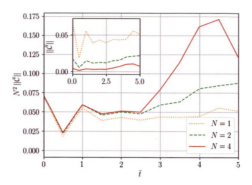

Figure B.2 Constraint violations $||\bar{\mathcal{C}}||$ multiplied by N^2 for the evolution of the electromagnetic wave initial data (B.55) with the reformulated Maxwell equations, using the script `maxwell.py`. We show results for $\bar{x}_{\text{out}} = 4$ and $n_{\text{grid}} = 2 + 16N$, i.e. with grid resolutions $\bar{\Delta} = 0.25/N$ (recall the use of two ghost zones). In the inset we show the constraint violations themselves; it can be seen that they do indeed get smaller with increasing resolution. At early times the rescaled constraint violations converge toward each other, demonstrating second-order convergence of the code. At later times, when the electromagnetic wave reaches the outer boundary, the code still converges but at a slower rate.

In writing our code, we adopted methods that are accurate only to a certain order and in a convergence test we verify that the errors decrease at the expected rate. In the script `maxwell.py` of Section B.3.2, for example, we adopted *second-order* finite differencing, meaning that the leading-order error terms should decrease with Δ^2, where Δ is the grid spacing (see, e.g., B.12 and B.13). If *all* errors scaled with Δ^2 then we could compute constraint violations, for example, for different resolutions, divide by Δ^2, and verify that the resulting rescaled constraint violations are always the same. In reality, the constraint violations will also be affected by higher-order error terms, meaning that the rescaled constraint violations will not be identical. As we increase the resolution, however, the higher-order error terms become smaller in comparison to the leading-order error term, meaning that the rescaled constraint violations will converge toward each other. This is shown in Fig. B.2, which demonstrates second-order convergence at early times. At later times the constraints are affected by approximations in the outer-boundary conditions (see the discussion around equation B.52), which results in a slower convergence.

Exercise B.9 Suggest and perform a test that verifies that the disappearance of second-order convergence around $\bar{t} \simeq 3$ in Fig. B.2 is indeed caused by the outer boundaries.

Readers interested in doing some Python coding, i.e. in modifying the script `maxwell.py`, may want to tackle the following exercise.

Exercise B.10 (a) Replace the second-order (iterative Crank–Nicholson) time integrator (B.47) implemented in `maxwell.py` (see the method `icn(self, fields, dt)`) with a new method that implements the fourth-order Runge–Kutta integrator (B.48).

(b) Replace the class `OperatorsOrder2` used in `maxwell.py` with a new class `OperatorsOrder4` that replaces second-order spatial differencing with fourth-order differencing (see Exercise B.2)). Note that centered fourth-order differencing requires two neighboring grid points on each side, so that this approach will now require two ghost zones at the boundaries rather than just one.

(c) Verify that your new code is fourth-order accurate at early times, until the evolution is affected by the outer boundaries.

We note that neither an analytical solution nor a constraint is needed to perform a convergence test, since we can always verify that a numerical code converges self-consistently. In this case we can run the code with at least four different resolutions and compute the difference between the results for "adjacent" pairs. These differences should again decrease at the rate expected from the order of the implementation. Rescaling the differences accordingly should result in a plot similar to Fig. B.2, where the convergence of the rescaled differences indicates the convergence of the code at the expected order.

Finally, we caution that a convergence test in the absence of an analytical solution provides a necessary, but not sufficient, test of a numerical code. It is always possible, for example, that the code contains an incorrect term or coefficient. Then simulations at successively higher resolutions might still converge at the expected order, but they would converge to the wrong answer.[11]

---

[11] We also note that singularities, as well as discontinuities in solutions or their derivatives, can lead to a reduction in the order of convergence, or, if the numerical scheme is poorly chosen, to convergence to an incorrect solution (see, e.g., Fig. 12.1 in LeVeque (1992) for an example).

Appendix C: A Very Brief Introduction to Matter Sources

In this book we have focused on vacuum spacetimes, and this has allowed us to treat black holes and gravitational waves. However, we recall that Einstein's equations have a non-zero stress–energy tensor T_{ab} on the right-hand side whenever matter sources are present. We included the matter source terms arising from the stress–energy tensor in our development of the 3+1 decomposition of Einstein's equations, as summarized in Boxes 2.1, 2.3, and 4.1. Also recall that the vanishing of the divergence of this stress–energy tensor, $\nabla_a T^{ab} = 0$, implied by the Bianchi identity (2.97), yields the equations of motion for these matter sources, which have to be solved together with Einstein's gravitational field equations whenever matter is present. These equations of motion are different for different matter sources, and we have not discussed them in this introductory exposition.

For hydrodynamics, which plays a crucial role in simulations of neutron stars and other astrophysical sources, the vanishing of the divergence of the fluid stress–energy tensor yields the equations of relativistic fluid dynamics. A discussion of this topic would go well beyond the scope of this book, hence we refer to other references for a detailed treatment of numerical techniques[1] and methods for solving these equations in the context of numerical relativity.[2] However, as we have already treated two matter fields in our development, i.e. scalar fields and electromagnetic fields, we will briefly sketch in this appendix how the equations of motion arise from the stress–energy tensor in these two cases.

---

[1] See, e.g., LeVeque (1992); Toro (1999).
[2] See, e.g., Baumgarte and Shapiro (2010); Rezzolla and Zanotti (2013); Shibata (2016).

C.1 Electromagnetic Fields

We will consider electromagnetic fields first. The stress–energy tensor T_{EM}^{ab} for these fields is given by[3]

$$T_{\text{EM}}^{ab} = \frac{1}{4\pi} \left(F^{ac} F^{b}_{\ c} - \frac{1}{4} g^{ab} F_{cd} F^{cd} \right), \tag{C.1}$$

where F^{ab} is the Faraday tensor that we encountered in Section 2.2.2, and where we write

$$F^{ab} = \nabla^a A^b - \nabla^b A^a \tag{C.2}$$

as in (2.25). We will assume that no other matter fields are present, for which we would have to include additional terms in the stress–energy tensor.

Taking the divergence of (C.1) we obtain

$$0 = \nabla_a T_{\text{EM}}^{ab} = \frac{1}{4\pi} \left(F^{b}_{\ c} \nabla_a F^{ac} + F^{ac} \nabla_a F^{b}_{\ c} - \frac{1}{2} g^{ab} F^{cd} \nabla_a F_{cd} \right), \tag{C.3}$$

where we have used $\nabla_a g^{ab} = 0$. We now rewrite the middle term on the right-hand side of (C.3) as follows:

$$F^{ac} \nabla_a F^{b}_{\ c} = g^{db} F^{ac} \nabla_a F_{dc} = \frac{1}{2} g^{db} F^{ac} \left(\nabla_a F_{dc} - \nabla_c F_{da} \right)$$

$$= \frac{1}{2} g^{ab} F^{dc} \left(\nabla_d F_{ac} - \nabla_c F_{ad} \right), \tag{C.4}$$

where we have used the antisymmetry of F^{ab} and have relabeled indices in the last equality. Inserting this into (C.3) yields

$$0 = F^{b}_{\ c} \nabla_a F^{ac} + \frac{1}{2} g^{ab} F^{cd} \left(-\nabla_d F_{ac} + \nabla_c F_{ad} - \nabla_a F_{cd} \right)$$

$$= F^{b}_{\ c} \nabla_a F^{ac} - \frac{1}{2} g^{ab} F^{cd} \left(\nabla_a F_{cd} + \nabla_d F_{ac} + \nabla_c F_{da} \right). \tag{C.5}$$

The last term, which is proportional to $\nabla_{[a} F_{cd]}$, vanishes identically when we insert (C.2), and this leaves us with

$$F^{b}_{\ c} \nabla_a F^{ac} = 0. \tag{C.6}$$

---

[3] This stress–energy tensor can be derived from the electromagnetic Lagrangian density $\mathcal{L} = -F^{ab} F_{ab}/(16\pi)$ with F^{ab} given by (C.2); see footnote 11 in Chapter 1.

In order to solve this equation for $\nabla_a F^{ac}$ we need to invert the tensor F^a_c, which is possible only when the determinant of its coefficients is non-zero.

> **Exercise C.1** (a) Evaluate the components F^a_b of the Faraday tensor (2.17) in a Lorentz frame, where $g_{ab} = \eta_{ab}$.
> (b) Show that $\det(F^a_b) = -(\mathbf{E} \cdot \mathbf{B})^2$.

In the generic case, when $\mathbf{E} \cdot \mathbf{B} \neq 0$,[4] we then have

$$\nabla_a F^{ab} = 0, \tag{C.7}$$

which is the Maxwell equation (2.19) in the absence of other matter fields. We see that the Bianchi identity, together with Einstein's equations, imply Maxwell's equations.

Adopting a 3+1 decomposition, we can write the stress–energy tensor (C.1) in terms of spatial and normal objects as

$$4\pi T_{\text{EM}}^{ab} = \frac{1}{2}(n^a n^b + \gamma^{ab})(E_i E^i + B_i B^i) + 2n^{(a}\epsilon^{b)cd}E_c B_d$$
$$- (E^a E^b + B^a B^b), \tag{C.8}$$

where E^a and B^a are the electric and magnetic fields observed by a normal observer, whose four-velocity is n^a, and where $\epsilon^{abc} = n_d \epsilon^{dabc}$. In particular we can now compute the source terms that appear in the 3+1 decomposition of Einstein's equations. Using equations (2.118), (2.120), and (2.124) we find

$$\rho_{\text{EM}} = n_a n_b T_{\text{EM}}^{ab} = \frac{1}{8\pi}\left(E_i E^i + B_i B^i\right)$$

$$= \frac{1}{8\pi}\left(\mathbf{E}^2 + \mathbf{B}^2\right) \qquad \text{(energy density)}, \tag{C.9}$$

$$j_i^{\text{EM}} = -\gamma_{ia} n_b T_{\text{EM}}^{ab} = \frac{1}{4\pi}\left(\epsilon_{ijk}E^j B^k\right)$$

$$= \frac{1}{4\pi}(\mathbf{E} \times \mathbf{B})_i \qquad \text{(Poynting vector)}, \tag{C.10}$$

---

[4] Special cases with $\mathbf{E} \cdot \mathbf{B} = 0$ admit counterexamples; see, e.g., p. 473 and exercise 20.8 in Misner et al. (2017).

$$S_{ij}^{\text{EM}} = \gamma_{ia}\gamma_{jb}T_{\text{EM}}^{ab}$$

$$= \frac{1}{8\pi}\left(\gamma_{ij}(E_k E^k + B_k B^k) - 2(E_i E_j + B_i B_j)\right) \quad \text{(stress tensor)},$$

$$\text{(C.11)}$$

$$S^{\text{EM}} = \gamma_{ij}S_{\text{EM}}^{ij} = \frac{1}{8\pi}\left(\mathbf{E}^2 + \mathbf{B}^2\right) \qquad \text{(trace of stress tensor).}$$

$$\text{(C.12)}$$

These results look familiar and should not surprise us. The source terms represent the energy density, energy flux, and stress as observed by the normal observer n^a; expressing these in terms of the electric and magnetic fields E^i and B^i observed by a normal observer therefore results in the usual energy density, Poynting flux, and stress for electromagnetic fields.

C.2 Scalar Fields

For scalar fields ψ, the stress–energy tensor is given by[5]

$$T_{ab}^{\psi} = (\nabla_a\psi)(\nabla_b\psi) - \frac{1}{2}g_{ab}\,g^{cd}(\nabla_c\psi)(\nabla_d\psi) - g_{ab}V(\psi), \quad \text{(C.13)}$$

where $V(\psi)$ is the potential. We may decompose the latter according to

$$V(\psi) = \frac{1}{2}m^2\psi^2 + V_{\text{int}}(\psi), \qquad \text{(C.14)}$$

where m is the mass of the field (i.e. the mass of the momentum eigenstates, or "particles" when the field is quantized), and $V_{\text{int}}(\psi)$ is an interaction potential. Here we have adopted "natural units", in which $c = 1$ and $\hbar = 1$.

Exercise C.2 (a) Show that the vanishing of the divergence $\nabla_a T_{\psi}^{ab} = 0$ yields the Klein–Gordon equation with a potential term,

$$\nabla_a\nabla^a\psi - m^2\psi - \frac{dV_{\text{int}}}{d\psi} = 0. \qquad \text{(C.15)}$$

(b) Define $\kappa \equiv -n^a\nabla_a\psi$ as in exercise 2.16, then decompose $\nabla_a\psi = n_a\kappa + D_a\psi$ (see equation 2.95) to show that the stress–energy tensor (C.13) may be written as

---

[5] This stress–energy tensor can be derived from the Lagrangian density
$\mathcal{L}_\psi = -g^{ab}\nabla_a\psi\nabla_b\psi/2 - V(\psi)$ for a scalar field ψ; see footnote 11 in Chapter 1.

$$T^{\psi}_{ab} = (n_a \kappa + D_a \psi)(n_b \kappa + D_b \psi) - \frac{1}{2}(\gamma_{ab} - n_a n_b)$$

$$\times \left((D_i \psi)(D^i \psi) - \kappa^2\right) - (\gamma_{ab} - n_a n_b)V(\psi). \tag{C.16}$$

(c) Evaluate equations (2.118), (2.120), and (2.124) to show that the source terms in the 3+1 decomposition of Einstein's equations are

$$\rho_\psi = n_a n_b T^{ab}_\psi$$

$$= \frac{1}{2}\left(\kappa^2 + (D_i \psi)(D^i \psi)\right) + V(\psi) \qquad \text{(energy density)}, \tag{C.17}$$

$$j^{\psi}_i = -\gamma_{ia} n_b T^{ab}_\psi = \kappa D_i \psi \qquad \text{(energy flux)}, \tag{C.18}$$

$$S^{\psi}_{ij} = \gamma_{ia}\gamma_{jb}T^{ab}_\psi$$

$$= (D_i \psi)(D_j \psi) - \frac{1}{2}\gamma_{ij}\left((D_k \psi)(D^k \psi) - \kappa^2\right)$$

$$- \gamma_{ij} V(\psi) \qquad \text{(stress tensor)}, \tag{C.19}$$

$$S^{\psi} = \gamma_{ij} S^{ij}_\psi = \frac{3}{2}\kappa^2 - \frac{1}{2}(D_k \psi)(D^k \psi) - 3V(\psi) \qquad \text{(trace of stress tensor)}. \tag{C.20}$$

Equation (C.15) is often invoked to describe the evolution of an "inflaton" field in inflationary cosmological models.[6] In a different context, a detailed study of black-hole formation with a scalar-field source has led to the discovery of "critical phenomena" in gravitational collapse.[7]

---

[6] See Starobinsky (1980); Guth (1981); Linde (1982); Albrecht and Steinhardt (1982); see also East et al. (2016); Clough et al. (2018); Giblin and Tishue (2019) for examples of some recent 3+1 numerical relativity simulations of inflationary cosmological models.
[7] See Choptuik (1993); see also Gundlach and Martín-García (2007) for a review.

Appendix D: A Summary of Important Results

As a reference for the reader we will collect in this appendix some important results and equations.

D.1 Differential Geometry

The results in this section hold in any number of dimensions. For spatial objects, simply restrict the indices to spatial indices only and replace the spacetime metric g_{ab} with the spatial metric γ_{ij} and the covariant derivative ∇_a with D_i. In order to be less dimensionally committed, we have also omitted superscripts (4) on objects like $^{(4)}R^a{}_{bcd}$ in this section.

- We denote the *metric* by g_{ab} and the inverse metric by g^{ab}. From the definition of the inverse metric we also have $g^a{}_b = g^{ac}g_{cb} = \delta^a{}_b$.
- In a coordinate basis, which we assume throughout this book, the *Christoffel symbols* can be computed from

$$\Gamma^a_{bc} = \frac{1}{2}g^{ad}(\partial_c g_{db} + \partial_b g_{dc} - \partial_d g_{bc}) \tag{D.1}$$

and are symmetric in the lower two indices,

$$\Gamma^a_{bc} = \Gamma^a_{cb}. \tag{D.2}$$

- The *covariant derivative* of a tensor involves a partial derivative and one Christoffel symbol Γ^a_{bc} for each index. A scalar does not have an index, therefore

$$\nabla_a \Phi = \partial_a \Phi. \tag{D.3}$$

For a contravariant rank-1 tensor (a "vector") we have

$$\nabla_a V^b = \partial_a V^b + V^c \Gamma^b_{ca}, \tag{D.4}$$

while for a covariant rank-1 tensor (a "one-form") we get

$$\nabla_a V_b = \partial_a V_b - V_c \Gamma^c_{ba}, \tag{D.5}$$

and for higher-rank tensors there is a Christoffel symbol term for each index, with the corresponding sign for contravariant or covariant indices, e.g.

$$\nabla_a T_b{}^c = \partial_a T_b{}^c - T_d{}^c \Gamma^d_{ba} + T_b{}^d \Gamma^c_{da}. \tag{D.6}$$

See also Box A.1 for some properties of the covariant derivative.

- The *Lie derivative* of a scalar function Φ along a vector **X** also reduces to a partial (directional) derivative,

$$\mathcal{L}_X \Phi = X^a \nabla_a \Phi = X^a \partial_a \Phi. \tag{D.7}$$

For a contravariant rank-1 tensor we have

$$\mathcal{L}_X v^a = X^b \nabla_b v^a - v^b \nabla_b X^a \tag{D.8}$$

(i.e. the commutator), while for a covariant rank-1 tensor we get

$$\mathcal{L}_X v_a = X^b \nabla_b v_a + v_b \nabla_a X^b. \tag{D.9}$$

The Lie derivative of a general tensor can again be constructed by including corresponding terms for all contravariant or covariant indices (see, e.g., p. 600 in Baumgarte and Shapiro (2010)). Covariant derivatives appearing in expressions for Lie derivatives can be replaced by ordinary partial derivatives.

- The *Riemann tensor* can be defined by

$$\nabla_a \nabla_b v_c - \nabla_b \nabla_a v_c = R^d_{cba} v_d \tag{D.10}$$

and computed from

$$R^a{}_{bcd} = \partial_c \Gamma^a_{bd} - \partial_d \Gamma^a_{bc} + \Gamma^a_{ec} \Gamma^e_{bd} - \Gamma^a_{ed} \Gamma^e_{bc}. \tag{D.11}$$

It satisfies the following symmetries:

$$R_{abcd} = -R_{bacd}, \quad R_{abcd} = -R_{abdc}, \quad R_{abcd} = R_{cdab}, \tag{D.12}$$
$$R_{abcd} + R_{adbc} + R_{acdb} = 0$$

as well as the *Bianchi identity*

$$\nabla_e R_{abcd} + \nabla_d R_{abec} + \nabla_c R_{abde} = 0. \tag{D.13}$$

In n dimensions the Riemann tensor has $n^2(n^2 - 1)/12$ independent components.

- The equation of *geodesic deviation* is

$$\frac{d^2 \Delta x^a}{d\tau^2} = -R^a{}_{bcd} u^b u^d \Delta x^c. \tag{D.14}$$

- The *Ricci tensor* and *Ricci scalar* are computed from contractions of the Riemann tensor,

$$R_{ab} \equiv R^c{}_{acb}, \qquad R \equiv g^{ab} R_{ab}. \tag{D.15}$$

Three common ways to compute the Ricci tensor are

$$R_{ab} = \partial_c \Gamma^c_{ab} - \partial_b \Gamma^c_{ac} + \Gamma^c_{ab} \Gamma^d_{cd} - \Gamma^c_{ad} \Gamma^d_{bc}, \tag{D.16}$$

$$R_{ab} = \frac{1}{2} g^{cd} \left(\partial_a \partial_d g_{cb} + \partial_c \partial_b g_{ad} - \partial_a \partial_b g_{cd} - \partial_c \partial_d g_{ab} \right)$$
$$+ g^{cd} \left(\Gamma^e_{ad} \Gamma_{ecb} - \Gamma^e_{ab} \Gamma_{ecd} \right), \tag{D.17}$$

$$R_{ab} = -\frac{1}{2} g^{cd} \partial_d \partial_c g_{ab} + g_{c(a} \partial_{b)} \Gamma^c + \Gamma^c \Gamma_{(ab)c} + 2 g^{ed} \Gamma^c_{e(a} \Gamma_{b)cd}$$
$$+ g^{cd} \Gamma^e_{ad} \Gamma_{ecb}, \tag{D.18}$$

where $\Gamma_{abc} = g_{ad} \Gamma^d_{bc}$ and $\Gamma^a \equiv g^{bc} \Gamma^a_{bc}$.

D.2 General Relativity

- The *Einstein tensor* is

$$G_{ab} \equiv {}^{(4)}R_{ab} - \frac{1}{2} g_{ab} {}^{(4)}R. \tag{D.19}$$

- *Einstein's equations* are

$$G_{ab} = 8\pi T_{ab} \tag{D.20}$$

where T_{ab} is the stress–energy tensor, and where we have assumed $\Lambda = 0$.

D.3 The 3+1 Decomposition

- A *foliation of spacetime* can be described by level surfaces of a function t. The *normal vector* to spatial surfaces, normalized so that $n_a n^a = -1$, is

$$n^a = -\alpha g^{ab} \nabla_b t \tag{D.21}$$

and can be expressed as

$$n_a = (-\alpha, 0, 0, 0) \quad \text{or} \quad n^a = \alpha^{-1}(1, -\beta^i). \tag{D.22}$$

Here α is the lapse function and β^i the shift vector.
- The induced or *spatial metric* is given by

$$\gamma_{ab} = g_{ab} + n_a n_b. \tag{D.23}$$

The spacetime metric can then be expressed as

$$g_{ab} = \begin{pmatrix} -\alpha^2 + \beta_k \beta^k & \beta_i \\ \beta_j & \gamma_{ij} \end{pmatrix} \tag{D.24}$$

or

$$g^{ab} = \begin{pmatrix} -\alpha^{-2} & \alpha^{-2}\beta^i \\ \alpha^{-2}\beta^j & \gamma^{ij} - \alpha^{-2}\beta^i\beta^j \end{pmatrix}. \tag{D.25}$$

- The *extrinsic curvature* is

$$K_{ab} \equiv -\gamma_a{}^c \gamma_b{}^d \nabla_c n_d = -\nabla_a n_b - n_a a_b = -\frac{1}{2}\mathcal{L}_{\mathbf{n}}\gamma_{ab}, \tag{D.26}$$

where $a_a \equiv n^b \nabla_b n_a$ is the acceleration of a normal observer.

D.4 The Equations of Gauss, Codazzi, Mainardi, and Ricci

- *Gauss's equation* relates the full spatial projection of the spacetime Riemann tensor ${}^{(4)}R^a_{bcd}$ to its spatial counterpart R^i_{jkl} as well as terms quadratic in the extrinsic curvature,

$$\gamma^p_i \gamma^q_j \gamma^r_k \gamma^s_l {}^{(4)}R_{pqrs} = R_{ijkl} + K_{ik}K_{jl} - K_{il}K_{jk}. \tag{D.27}$$

- The *Codazzi equation* (sometimes called the *Codazzi–Mainardi relation*) relates three spatial and one normal projection of the spacetime Riemann tensor to spatial derivatives of the extrinsic curvature,

$$\gamma^p_i \gamma^q_j \gamma^r_k n^s {}^{(4)}R_{pqrs} = D_j K_{ik} - D_i K_{jk}. \tag{D.28}$$

- Finally, *Ricci's equation* relates two spatial and two normal projections of the spacetime Riemann tensor to the Lie derivative of the extrinsic curvature along the normal vector n^a, plus other spatial terms,

$$\gamma_i^p \gamma_j^q n^r n^{s\,(4)} R_{prqs} = \mathcal{L}_n K_{ij} + \frac{1}{\alpha} D_i D_j \alpha + K_i^k K_{jk}. \qquad (D.29)$$

D.5 The ADM Equations

- The *ADM equations* consist of the *constraint equations*

$$R + K^2 - K_{ij} K^{ij} = 16\pi\rho \quad \text{(Hamiltonian constraint)}, \qquad (D.30)$$

$$D_i(K^{ij} - \gamma^{ij} K) = 8\pi S^i \quad \text{(momentum constraint)} \qquad (D.31)$$

and the *evolution equations*

$$\partial_t \gamma_{ij} = -2\alpha K_{ij} + D_i \beta_j + D_j \beta_i, \qquad (D.32)$$

$$\partial_t K_{ij} = \alpha(R_{ij} - 2K_{ik} K^k{}_j + K K_{ij}) - D_i D_j \alpha$$

$$- 8\pi\alpha \left(S_{ij} - \frac{1}{2}\gamma_{ij}(S - \rho) \right) + \beta^k \partial_k K_{ij} + K_{ik} \partial_j \beta^k + K_{kj} \partial_i \beta^k. \qquad (D.33)$$

D.6 Conformal Decompositions

- In a conformal decomposition of the spatial metric we write

$$\gamma_{ij} = \psi^4 \bar{\gamma}_{ij}, \qquad (D.34)$$

where ψ is the *conformal factor* and $\bar{\gamma}_{ij}$ is the *conformally related metric*.

- In the context of the constraint equations we often decompose the extrinsic curvature according to

$$K_{ij} = A_{ij} + \frac{1}{3}\gamma_{ij} K = \psi^{-2} \bar{A}_{ij} + \frac{1}{3}\psi^4 \bar{\gamma}_{ij} K \qquad (D.35)$$

(see equation 4.26 for a common decomposition in the context of the evolution equations).

- The Hamiltonian constraint then becomes

$$\bar{D}^2 \psi - \frac{\psi}{8}\bar{R} + \frac{\psi^{-7}}{8}\bar{A}_{ij}\bar{A}^{ij} - \frac{1}{12}\psi^5 K^2 = -2\pi\psi^5\rho, \qquad (D.36)$$

where $\bar{D}$ and $\bar{R}$ are the covariant derivative and the Ricci scalar associated with $\bar{\gamma}_{ij}$, and the momentum constraint becomes

$$\bar{D}_j \bar{A}^{ij} - \frac{2}{3}\psi^6 \bar{\gamma}^{ij} \bar{D}_j K = 8\pi\psi^{10} S^i. \qquad (D.37)$$

Appendix E: Answers to Selected Problems

Exercise 1.11: (c) $R(t) = 4\left((\mu M^2/5)(T - t)\right)^{1/4}$.

Exercise 2.3: (a) You should find $n_{a'} = -\alpha(1, - h', 0, 0)$ and $n^{a'} = \alpha(1, h', 0, 0)$, where $h' = dh/dr$ (not to be confused with a primed index).

(b) From $n_{a'}n^{a'} = \alpha^2(-1+(h')^2)$ you should find $\alpha = (1-(h')^2)^{-1/2}$.

(c) Evaluate $n_a = (\partial x^{b'}/\partial x^a) n_{b'}$. For $a = t$ the only non-zero term is $n_t = (\partial T/\partial t) n_T = -\alpha$; for $a = r$ we have $n_r = (\partial T/\partial r) n_T + (\partial r/\partial r) n_r = (h')(-\alpha) + \alpha h' = 0$. Therefore $n_a = (-\alpha, 0, 0, 0)$, as expected.

Exercise 2.4: Now evaluate $n^a = (\partial x^a/\partial x^{b'}) n^{b'}$ to find $n^a = \alpha^{-1}(1, \alpha^2 h', 0, 0)$. Comparing with (2.40) we identify $\beta^i = (-\alpha^2 h', 0, 0)$.

Exercise 2.6: Evaluate $\gamma^a_b = g^a_b + n^a n_b = \delta^a_b + n^a n_b$ to find $\gamma^t_t = 0$, $\gamma^r_t = -\alpha^2 h'$, $\gamma^t_r = 0$, as well as $\gamma^r_r = \gamma^\theta_\theta = \gamma^\varphi_\varphi = 1$; all other components vanish.

Exercise 2.8: (c) The other two non-zero Christoffel symbols are $\Gamma^t_{rr} = h''$ and $\Gamma^t_{\varphi\varphi} = r \sin^2\theta\, h'$.

(d) You should find $K_{rr} = -\alpha h''$, $K_{\theta\theta} = -\alpha r h'$, and $K_{\varphi\varphi} = \sin^2\theta\, K_{\theta\theta}$.

Exercise 2.20: (a) You should find $\beta_{\bar{r}} = M/\bar{r}$ and $\gamma_{ij} = \psi^4 \eta_{ij}$, with $\psi = (1 + M/\bar{r})^{1/2}$.

(b) $\gamma^{ij} = \psi^{-4}\eta^{ij}$, and $\beta^{\bar{r}} = \bar{r}M/(\bar{r} + M)^2$.

(c) $\alpha = \bar{r}/(\bar{r} + M)$.

Exercise 2.21: The lapse and shift should be as before; the spatial metric is $\gamma_{ij} = \mathrm{diag}(1 - (h')^2, r^2, r^2 \sin^2\theta)$.

Exercise 2.24: (a) The non-zero Christoffel symbols are

$$\Gamma^r_{rr} = -\alpha^2 h' h'', \qquad\qquad \Gamma^r_{\theta\theta} = -\alpha^2 r,$$
$$\Gamma^r_{\varphi\varphi} = -\alpha^2 r \sin^2\theta, \qquad \Gamma^\theta_{\theta r} = r^{-1},$$
$$\Gamma^\theta_{\varphi\varphi} = -\sin\theta\cos\theta, \qquad \Gamma^\varphi_{\varphi r} = r^{-1}, \qquad\qquad \text{(E.1)}$$
$$\Gamma^\varphi_{\theta\varphi} = -\cot\theta,$$

where $\alpha = (1 - (h')^2)^{-1/2}$. Note that for $h = const$, and hence $\alpha = 1$, these reduce to the spatial Christoffel symbols in spherical coordinates; see equation (A.47).

(b) The non-zero components of the Ricci tensor are $R_{rr} = -2\alpha^2 h' h''/r$, $R_{\theta\theta} = -\alpha^2 (h')^2 - r\alpha^4 h' h''$, and $R_{\varphi\varphi} = R_{\theta\theta}\sin^2\theta$.

Exercise 2.25: First note that the inverse of the spatial metric found in exercise 2.21 can be written as $\gamma^{ij} = \text{diag}(\alpha^2, r^{-2}, r^{-2}\sin^{-2}\theta)$. Then contract this metric with the Ricci tensor of exercise 2.24 to find $R = \gamma^{ij} R_{ij} = -4\alpha^4 h' h''/r - 2\alpha^2 (h')^2/r^2$. From the extrinsic curvature of exercise 2.8 we similarly find $K = \gamma^{ij} K_{ij} = -\alpha^3 h'' - 2\alpha h'/r$ and therefore $K^2 = \alpha^6 (h'')^2 + 4\alpha^4 h' h''/r + 4\alpha^2 (h')^2/r^2$. Finally we compute $K_{ij}K^{ij} = \alpha^6 (h'')^2 + 2\alpha^2 (h')^2/r^2$. Combining these we have $R + K^2 - K_{ij}K^{ij} = 0$, as expected.

Exercise 2.27: (a) In addition to $K_{\bar{r}\bar{r}} = -M/\bar{r}^2$ you should find $K_{\theta\theta} = M$ and $K_{\varphi\varphi} = \sin^2\theta\, K_{\theta\theta}$.

(b) Either way you should find $K = M/(\bar{r} + M)^2$.

Exercise 3.2: Using the results of exercises 2.20 and 2.27 you should find $K^2 = M^2/(\bar{r} + M)^4 = \psi^{-8} M^2/\bar{r}^4$ and $K_{ij}K^{ij} = 3\psi^{-8} M^2/\bar{r}^4$. Also compute

$$\bar{D}^2\psi = \frac{1}{\bar{r}^2}\frac{d}{d\bar{r}}\left(\bar{r}^2\frac{d\psi}{d\bar{r}}\right) = -\psi^{-3}\frac{M^2}{4\bar{r}^4}. \qquad\qquad \text{(E.2)}$$

Inserting these into the Hamiltonian constraint (3.12) shows that all terms cancel each other, as they should.

Exercise 4.5: $f(\alpha) = (1 - \alpha)/\alpha$.

Exercise B.2: You should find $f'_i = (-f_{i+2} + 8f_{i+1} - 8f_{i-1} + f_{i-2})/(12\Delta) + \mathcal{O}(\Delta^4)$ and $f''_i = (-f_{i+2} + 16 f_{i+1} - 30 f_i + 16 f_{i-1} - f_{i-2})/(12\Delta^2) + \mathcal{O}(\Delta^4)$.

References

Aasi, J. et al. (The LIGO, VIRGO and NINJA-2 Collaborations) (2014). The NINJA-2 project: Detecting and characterizing gravitational waveforms modeled using numerical binary black hole simulations. *Class. Quantum Grav. 31*, 115004.

Abbott et al. (LIGO Scientific Collaboration and Virgo Collaboration) (2016a). Observation of gravitational waves from a binary black hole merger. *Phys. Rev. Lett. 116*, 061102.

Abbott et al. (LIGO Scientific Collaboration and Virgo Collaboration) (2016b). Properties of the binary black hole merger GW150914. *Phys. Rev. Lett. 116*, 241102.

Abbott et al. (LIGO Scientific Collaboration and Virgo Collaboration) (2017). GW170817: Observation of gravitational waves from a binary neutron star inspiral. *Phys. Rev. Lett. 119*, 161101.

Abbott et al. (LIGO Scientific Collaboration and Virgo Collaboration) (2018). GWTC-1: A gravitational-wave transient catalog of compact binary mergers observed by LIGO and Virgo during the first and second observing runs. *Phys. Rev. X 9*, 031040.

Ajith, P., S. Babak, Y. Chen, M. Hewitson, B. Krishnan, A. M. Sintes, J. T. Whelan, B. Brügmann, P. Diener, N. Dorband, J. Gonzalez, M. Hannam, S. Husa, D. Pollney, L. Rezzolla, L. Santamaría, U. Sperhake, and J. Thornburg (2008). Template bank for gravitational waveforms from coalescing binary black holes: Nonspinning binaries. *Phys. Rev. D 77*, 104017.

Ajith, P., M. Boyle, D. A. Brown, B. Brügmann, L. T. Buchman, L. Cadonati, M. Campanelli, T. Chu, Z. B. Etienne, S. Fairhurst, M. Hannam, J. Healy, I. Hinder, S. Husa, L. E. Kidder, B. Krishnan, P. Laguna, Y. T. Liu, L. London, C. O. Lousto, G. Lovelace, I. MacDonald, P. Marronetti, S. Mohapatra, P. Mösta, D. Müller, B. C. Mundim, H. Nakano, F. Ohme, V. Paschalidis, L. Pekowsky, D. Pollney, H. P. Pfeiffer, M. Ponce, M. Pürrer, G. Reifenberger, C. Reisswig, L. Santamaría, M. A. Scheel, S. L. Shapiro, D. Shoemaker, C. F. Sopuerta, U. Sperhake, B. Szilágyi, N. W. Taylor, W. Tichy, P. Tsatsin, and Y. Zlochower (2012). The NINJA-2 catalog of hybrid post-Newtonian/numerical-relativity waveforms for non-precessing black-hole binaries. *Class. Quantum Grav. 29*, 124001.

Ajith, P., M. Hannam, S. Husa, Y. Chen, B. Brügmann, N. Dorband, D. Müller, F. Ohme, D. Pollney, C. Reisswig, L. Santamaría, and J. Seiler (2011). Inspiral-merger-ringdown waveforms for black-hole binaries with nonprecessing spins. *Phys. Rev. Lett. 106*, 241101.

Albrecht, A. and P. J. Steinhardt (1982). Cosmology for grand unified theories with radiatively induced symmetry breaking. *Phys. Rev. Lett. 48*, 1220.

Alcubierre, M. (2008). *Introduction to 3+1 Numerical Relativity*. Oxford University Press, New York.

Alcubierre, M. and B. Brügmann (2001). Simple excision of a black hole in 3+1 numerical relativity. *Phys. Rev. D 63*, 104006.

Alcubierre, M., B. Brügmann, P. Diener, M. Koppitz, D. Pollney, E. Seidel, and R. Takahashi (2003). Gauge conditions for long-term numerical black hole evolutions without excision. *Phys. Rev. D 67*, 084023.

Alic, D., C. Bona-Casas, C. Bona, L. Rezzolla, and C. Palenzuela (2012). Conformal and covariant formulation of the Z4 system with constraint-violation damping. *Phys. Rev. D 85*, 064040.

Ansorg, M., B. Brügmann, and W. Tichy (2004). Single-domain spectral method for black hole puncture data. *Phys. Rev. D 70*, 064011.

Arnowitt, R., S. Deser, and C. W. Misner (1962). The dynamics of general relativity. In L. Witten (ed.), *Gravitation: An Introduction to Current Research*, pp. 227. Wiley, New York. Also available at gr-qc/0405109.

Aylott, B., J. G. Baker, W. D. Boggs, M. Boyle, P. R. Brady, D. A. Brown, B. Brügmann, L. T. Buchman, A. Buonanno, L. Cadonati, J. Camp, M. Campanelli, J. Centrella, S. Chatterji, N. Christensen, T. Chu, P. Diener, N. Dorband, Z. B. Etienne, J. Faber, S. Fairhurst, B. Farr, S. Fischetti, G. Guidi, L. M. Goggin, M. Hannam, F. Herrmann, I. Hinder, S. Husa, V. Kalogera, D. Keppel, L. E. Kidder, B. J. Kelly, B. Krishnan, P. Laguna, C. O. Lousto, I. Mandel, P. Marronetti, R. Matzner, S. T. McWilliams, K. D. Matthews, R. A. Mercer, S. R. P. Mohapatra, A. H. Mroué, H. Nakano, E. Ochsner, Y. Pan, L. Pekowsky, H. P. Pfeiffer, D. Pollney, F. Pretorius, V. Raymond, C. Reisswig, L. Rezzolla, O. Rinne, C. Robinson, C. Röver, L. Santamaría, B. Sathyaprakash, M. A. Scheel, E. Schnetter, J. Seiler, S. L. Shapiro, D. Shoemaker, U. Sperhake, A. Stroeer, R. Sturani, W. Tichy, Y. T. Liu, M. van der Sluys, J. R. van Meter, R. Vaulin, A. Vecchio, J. Veitch, A. Viceré, J. T. Whelan, and Y. Zlochower (2009). Testing gravitational-wave searches with numerical relativity waveforms: Results from the first Numerical INJection Analysis (NINJA) project. *Class. Quantum Grav. 26*, 165008.

Baker, J. G., M. Campanelli, F. Pretorius, and Y. Zlochower (2007). Comparisons of binary black hole merger waveforms. *Class. Quantum Grav. 24*, S25.

Baker, J. G., J. Centrella, D.-I. Choi, M. Koppitz, and J. van Meter (2006). Gravitational-wave extraction from an inspiraling configuration of merging black holes. *Phys. Rev. Lett. 96*, 111102.

Baumgarte, T. W. (2000). Innermost stable circular orbit of binary black holes. *Phys. Rev. D 62*, 024018.

Baumgarte, T. W. (2012). Alternative approach to solving the Hamiltonian constraint. *Phys. Rev. D 85*, 084013.

Baumgarte, T. W. and S. G. Naculich (2007). Analytical representation of a black hole puncture solution. *Phys. Rev. D 75*, 067502.

Baumgarte, T. W. and S. L. Shapiro (1998). Numerical integration of Einstein's field equations. *Phys. Rev. D 59*, 024007.

Baumgarte, T. W. and S. L. Shapiro (2010). *Numerical Relativity: Solving Einstein's Equations on the Computer*. Cambridge University Press, Cambridge.

Baumgarte, T. W., P. J. Montero, I. Cordero-Carrión, and E. Müller (2013). Numerical relativity in spherical polar coordinates: Evolution calculations with the BSSN formulation. *Phys. Rev. D 87*, 044026.

Beig, R. and N. Ó. Murchadha (1994). Trapped surfaces in vacuum spacetimes. *Class. Quantum Grav. 11*, 419.

Beig, R. and N. Ó. Murchadha (1996). Vacuum spacetimes with future trapped surfaces. *Class. Quantum Grav. 13*, 739.

Beig, R. and N. Ó. Murchadha (1998). Late time behavior of the maximal slicing of the Schwarzschild black hole. *Phys. Rev. D 57*, 4728.

Bergmann, P. G. (1957). Summary of the Chapel Hill Conference. *Rev. Mod. Phys. 29*, 352.

Bernuzzi, S. and D. Hilditch (2010). Constraint violation in free evolution schemes: Comparing the BSSNOK formulation with a conformal decomposition of the Z4 formulation. *Phys. Rev. D 81*, 084003.

Blecha, L., D. Sijacki, L. Z. Kelley, P. Torrey, M. Vogelsberger, D. Nelson, V. Springel, G. Snyder, and L. Hernquist (2016). Recoiling black holes: Prospects for detection and implications of spin alignment. *Mon. Not. Roy. Astron. Soc. 456*, 961.

Bogdanovic, T., C. S. Reynolds, and M. C. Miller (2007). Alignment of the spins of supermassive black holes prior to merger. *Astrophys. J. Lett. 661*, L147.

Bona, C., T. Ledvinka, C. Palenzuela, and M. Zacek (2003). General-covariant evolution formalism for numerical relativity. *Phys. Rev. D 67*, 104005.

Bona, C., J. Massó, E. Seidel, and J. Stela (1995). New formalism for numerical relativity. *Phys. Rev. Lett. 75*, 600.

Bona, C. and C. Palenzuela (2005). *Elements of Numerical Relativity: From Einstein's Equations to Black Hole Simulations*. Springer, Berlin, Heidelberg.

Bona, C., C. Palenzuela-Luque, and C. Bona-Casas (2009). *Elements of Numerical Relativity and Relativistic Hydrodynamics*. Springer, Berlin, Heidelberg.

Bonazzola, S., E. Gourgoulhon, P. Grandclément, and J. Novak (2004). A constrained scheme for Einstein equations based on Dirac gauge and spherical coordinates. *Phys. Rev. D 70*, 104007.

Bowen, J. M. and J. W. York, Jr. (1980). Time-asymmetric initial data for black holes and black hole collisions. *Phys. Rev. D 21*, 2047.

Brandt, S. and B. Brügmann (1997). A simple construction of initial data for multiple black holes. *Phys. Rev. Lett. 78*, 3606.

Brill, D. R. and R. W. Lindquist (1963). Interaction energy in geometrostatics. *Phys. Rev. 131*, 471.

Brown, J. D. (2008). Puncture evolution of Schwarzschild black holes. *Phys. Rev. D 77*, 044018.

Brown, J. D. (2009). Covariant formulations of Baumgarte, Shapiro, Shibata and Nakamura and the standard gauge. *Phys. Rev. D 79*, 104029.

Brügmann, B., W. Tichy, and N. Jansen (2004). Numerical simulation of orbiting black holes. *Phys. Rev. Lett. 92*, 211101.

Buonanno, A., G. B. Cook, and F. Pretorius (2007). Inspiral, merger and ring-down of equal-mass black-hole binaries. *Phys. Rev. D 75*, 124018.

Buonanno, A. and T. Damour (1999). Effective one-body approach to general relativistic two-body dynamics. *Phys. Rev. D 59*, 084006.

Campanelli, M., C. O. Lousto, P. Marronetti, and Y. Zlochower (2006a). Accurate evolutions of orbiting black-hole binaries without excision. *Phys. Rev. Lett. 96*, 111101.

Campanelli, M., C. O. Lousto, and Y. Zlochower (2006b). Spinning black holes: The orbital hang-up. *Phys. Rev. D 74*, 041501.

Campanelli, M., C. O. Lousto, Y. Zlochower, and D. Merritt (2007a). Large merger recoils and spin flips from generic black-hole binaries. *Astrophys. J. Lett. 659*, L5.

Campanelli, M., C. O. Lousto, Y. Zlochower, and D. Merritt (2007b). Maximum gravitational recoil. *Phys. Rev. Lett. 98*, 231102.

Cardoso, V., L. Gualtieri, C. A. R. Herdeiro, and U. Sperhake (2015). Exploring new physics frontiers through numerical relativity. *Living Rev. Rel. 18*, 1.

Carroll, S. (2004). *Spacetime and Geometry: An Introduction to General Relativity.* Addison-Wesley, San Francisco. Reissued in 2019 by Cambridge University Press.

Caudill, M., G. B. Cook, J. D. Grigsby, and H. P. Pfeiffer (2006). Circular orbits and spin in black-hole initial data. *Phys. Rev. D 74*, 064011.

Chiaberge, M., G. R. Tremblay, A. Capetti, and C. Norman (2018). The recoiling black hole candidate 3C 186: Spatially resolved quasar feedback and further evidence of a blueshifted broad-line region. *Astrophys. J. 861*, 56.

Choptuik, M. W. (1993). Universality and scaling in gravitational collapse of a massless scalar field. *Phys. Rev. Lett. 70*, 9.

Choquet-Bruhat, Y. (2015). *Introduction to General Relativity, Black Holes and Cosmology.* Oxford University Press, Oxford.

Clough, K., R. Flauger, and E. A. Lim (2018). Robustness of inflation to large tensor perturbations. *J. Cosmology Astroparticle Phys. 2018*, 065.

Cook, G. B. (1991). Initial data for axisymmetric black-hole collisions. *Phys. Rev. D 44*, 2983.

Cook, G. B. (1994). Three-dimensional initial data for the collision of two black holes. II. Quasicircular orbits for equal-mass black holes. *Phys. Rev. D 50*, 5025.

Cook, G. B. and H. P. Pfeiffer (2004). Excision boundary conditions for black-hole initial data. *Phys. Rev. D 70*, 104016.

De Donder, T. (1921). *La gravifique einsteinienne.* Gauthier-Villars, Paris.

de Felice, A., F. Larrouturou, S. Mukohyama, and M. Oliosi (2019). On the absence of conformally flat slicings of the Kerr spacetime. *Phys. Rev. D 100*, 124044.

Dennison, K. A. and T. W. Baumgarte (2014). A simple family of analytical trumpet slices of the Schwarzschild spacetime. *Class. Quantum Grav. 31*, 117001.

Dennison, K. A., T. W. Baumgarte, and P. J. Montero (2014). Trumpet slices in Kerr spacetimes. *Phys. Rev. Lett. 113*, 261101.

Droste, J. (1917). The field of a single centre in Einstein's theory of gravitation, and the motion of a particle in that field. *Koninklijke Nederlandse Akademie van Wetenschappen Proceedings, Series B Phys. Sci. 19*, 197.

East, W. E., M. Kleban, A. Linde, and L. Senatore (2016). Beginning inflation in an inhomogeneous universe. *J. Cosmology Astroparticle Phys. 2016*, 010.

Einstein, A. (1915). Die Feldgleichungen der Gravitation. *Preuss. Akad. Wiss. Berlin, Sitzber.*, 844.

Einstein, A. (1916). Näherungsweise Integration der Feldgleichungen der Gravitation. *Preuss. Akad. Wiss. Berlin, Sitzber.*, 688.

Estabrook, F., H. Wahlquist, S. Christensen, B. Dewitt, L. Smarr, and E. Tsiang (1973). Maximally slicing a black hole. *Phys. Rev. D 7*, 2814.

Etienne, Z. B., Y. T. Liu, and S. L. Shapiro (2010). Relativistic magnetohydrodynamics in dynamical spacetimes: A new adaptive mesh refinement implementation. *Phys. Rev. D 82*, 084031.

Etienne, Z. B., V. Paschalidis, R. Haas, P. Mösta, and S. L. Shapiro (2015). Illinois-GRMHD: An open-source, user-friendly GRMHD code for dynamical spacetimes. *Class. Quantum Grav. 32*, 175009.

Etienne, Z. B., M.-B. Wan, M. C. Babiuc, S. T. McWilliams, and A. Choudhary (2017). GiRaFFE: An open-source general relativistic force-free electrodynamics code. *Class. Quantum Grav. 34*, 215001.

Friedrich, H. (1985). On the hyperbolicity of Einstein's and other gauge field equations. *Commun. Math. Phys. 100*, 525.

Garat, A. and R. H. Price (2000). Nonexistence of conformally flat slices of the Kerr spacetime. *Phys. Rev. D 61*, 124011.

Garfinkle, D. (2002). Harmonic coordinate method for simulating generic singularities. *Phys. Rev. D 65*, 044029.

Gerosa, D., M. Kesden, R. O'Shaughnessy, A. Klein, E. Berti, U. Sperhake, and D. Trifirò (2015). Precessional instability in binary black holes with aligned spins. *Phys. Rev. Lett. 115*, 141102.

Giblin, J. T. and A. J. Tishue (2019). Preheating in full general relativity. *Phys. Rev. D 100*, 063543.

Gieg, H., T. Dietrich, and M. Ujevic (2019). Simulating binary neutron stars with hybrid equation of states: Gravitational waves, electromagnetic signatures, and challenges for numerical relativity. *arXiv e-prints*, arXiv:1908.03135.

Gleiser, R. J., C. O. Nicasio, R. H. Price, and J. Pullin (1998). Evolving the Bowen–York initial data for spinning black holes. *Phys. Rev. D 57*, 3401.

Gleiser, R. J., G. Khanna, and J. Pullin (2002). Evolving the Bowen–York initial data for boosted black holes. *Phys. Rev. D 66*, 024035.

González, J. A., M. D. Hannam, U. Sperhake, B. Brugmann, and S. Husa (2007). Supermassive recoil velocities for binary black-hole mergers with antialigned spins. *Phys. Rev. Lett. 98*, 231101.

González, J. A., U. Sperhake, B. Brügmannn, M. Hannam, and S. Husa (2007). Total recoil: The maximum kick from nonspinning black-hole binary inspiral. *Phys. Rev. Lett. 98*, 091101.

Gopal-Krishna, P. L. Biermann, L. Á. Gergely, and P. J. Wiita (2012). On the origin of X-shaped radio galaxies. *Research in Astronomy and Astrophysics 12*, 127.

Gourgoulhon, E. (2012). *3+1 Formalism in General Relativity*. Springer, New York.

Gourgoulhon, E., P. Grandclément, and S. Bonazzola (2002). Binary black holes in circular orbits. I. A global spacetime approach. *Phys. Rev. D 65*, 044020.

Grandclément, P., E. Gourgoulhon, and S. Bonazzola (2002). Binary black holes in circular orbits. II. Numerical methods and first results. *Phys. Rev. D 65*, 044021.

Griffiths, D. J. (2013). *Introduction to Electrodynamics* (4th edn). Cambridge University Press, Cambridge.

Gullstrand, A. (1922). Allgemeine Lösung des statischen Einkörperproblems in der Einsteinschen Gravitationstheorie. *Arkiv för Matematik, Astronomi och Fysik 16*, 1.

Gundlach, C. and J. M. Martín-García (2007). Critical phenomena in gravitational collapse. *Living Rev. Rel. 10*, 5.

Gundlach, C., J. M. Martín-García, G. Calabrese, and I. Hinder (2005). Constraint damping in the Z4 formulation and harmonic gauge. *Class. Quantum Grav. 22*, 3767.

Guth, A. H. (1981). Inflationary universe: A possible solution to the horizon and flatness problems. *Phys. Rev. D 23*, 347.

Hahn, S. G. and R. W. Lindquist (1964). The two-body problem in geometrodynamics. *Ann. Phys. 29*, 304.

Hannam, M., S. Husa, N. Ó. Murchadha, B. Brügmann, J. A. González, and U. Sperhake (2007a). Where do moving punctures go? *J. Phys. Conf. Series 66*, 012047.

Hannam, M., S. Husa, D. Pollney, B. Brügmann, and N. O'Murchadha (2007b). Geometry and regularity of moving punctures. *Phys. Rev. Lett. 99*, 241102.

Hannam, M., S. Husa, F. Ohme, B. Brügmann, and N. Ó. Murchadha (2008). Wormholes and trumpets: The Schwarzschild spacetime for the moving-puncture generation. *Phys. Rev. D 78*, 064020.

Hartle, J. B. (2003). *Gravity: An Introduction to Einstein's General Relativity*. Addison Wesley, San Francisco.

Hawking, S. W. and G. F. R. Ellis (1973). *The Large Scale Structure of Space-Time*. Cambridge University Press, Cambridge.

Healy, J. and C. O. Lousto (2018). Hangup effect in unequal mass binary black hole mergers and further studies of their gravitational radiation and remnant properties. *Phys. Rev. D 97*, 084002.

Healy, J., J. Lange, R. O'Shaughnessy, C. O. Lousto, M. Campanelli, A. R. Williamson, Y. Zlochower, J. Calderón Bustillo, J. A. Clark, C. Evans, D. Ferguson, S. Ghonge, K. Jani, B. Khamesra, P. Laguna, D. M. Shoemaker, M. Boyle, A. García, D. A. Hemberger, L. E. Kidder, P. Kumar, G. Lovelace, H. P. Pfeiffer, M. A. Scheel, and S. A. Teukolsky (2018). Targeted numerical simulations of binary black holes for GW170104. *Phys. Rev. D 97*, 064027.

Healy, J., C. O. Lousto, J. Lange, R. O'Shaughnessy, Y. Zlochower, and M. Campanelli (2019). The second RIT binary black hole simulations catalog and its application to gravitational waves parameter estimation. *Phys. Rev. D 100*, 024021.

Herrmann, F., I. Hinder, D. Shoemaker, and P. Laguna (2007). Unequal-mass binary black hole plunges and gravitational recoil. *Class. Quantum Grav. 24*, S33.

Hilbert, D. (1915). Grundlagen der Physik. *Gott. Nachr. 27*, 395.

Hilditch, D. (2015). Dual foliation formulations of general relativity. *arXiv e-prints*, arXiv:1509.02071.

Hinder, I., A. Buonanno, M. Boyle, Z. B. Etienne, J. Healy, N. K. Johnson-McDaniel, A. Nagar, H. Nakano, Y. Pan, H. P. Pfeiffer, M. Pürrer, C. Reisswig, M. A. Scheel, E. Schnetter, U. Sperhake, B. Szilágyi, W. Tichy, B. Wardell, A. Zenginoğlu, D. Alic, S. Bernuzzi, T. Bode, B. Brügmann, L. T. Buchman, M. Campanelli, T. Chu, T. Damour, J. D. Grigsby, M. Hannam, R. Haas, D. A. Hemberger,

S. Husa, L. E. Kidder, P. Laguna, L. London, G. Lovelace, C. O. Lousto, P. Marronetti, R. A. Matzner, P. Mösta, A. Mroué, D. Müller, B. C. Mundim, A. Nerozzi, V. Paschalidis, D. Pollney, G. Reifenberger, L. Rezzolla, S. L. Shapiro, D. Shoemaker, A. Taracchini, N. W. Taylor, S. A. Teukolsky, M. Thierfelder, H. Witek, and Y. Zlochower (2013). Error-analysis and comparison to analytical models of numerical waveforms produced by the NRAR Collaboration. *Class. Quantum Grav. 31*, 025012.

Hulse, R. A. and J. H. Taylor (1975). Discovery of a pulsar in a binary system. *Astrophys. J. Lett. 195*, L51.

Isenberg, J. (1978). Unpublished.

Jackson, J. D. (1999). *Classical Electrodynamics* (3rd edn). Wiley, New York.

Kennefick, D. (2005). Einstein versus the *Physical Review*. *Physics Today 58*, 43.

Kerr, R. P. (1963). Gravitational field of a spinning mass as an example of algebraically special metrics. *Phys. Rev. Lett. 11*, 237.

Khan, S., K. Chatziioannou, M. Hannam, and F. Ohme (2018). Phenomenological model for the gravitational-wave signal from precessing binary black holes with two-spin effects. *Phys. Rev. D 100*, 024059.

Kidder, L. E., S. E. Field, F. Foucart, E. Schnetter, S. A. Teukolsky, A. Bohn, N. Deppe, P. Diener, F. Hébert, J. Lippuner, J. Miller, C. D. Ott, M. A. Scheel, and T. Vincent (2017). SpECTRE: A task-based discontinuous Galerkin code for relativistic astrophysics. *J. Comp. Phys. 335*, 84.

Knapp, A. M., E. J. Walker, and T. W. Baumgarte (2002). Illustrating stability properties of numerical relativity in electrodynamics. *Phys. Rev. D 65*, 064031.

Komossa, S. (2012). Recoiling black holes: Electromagnetic signatures, candidates, and astrophysical implications. *Adv. Astron. 2012*, 364973.

Lanczos, C. (1922). Ein vereinfachtes Koordinatensystem für die Einsteinschen Gravitationsgleichungen. *Phys. Z. 23*, 537.

LeVeque, R. (1992). *Numerical Methods for Conservation Laws* (2nd edn). Birkhäuser, Basel, Switzerland; Boston, USA.

Lichnerowicz, A. (1944). L'intégration des équations de la gravitation relativiste et la problème des *n* corps. *J. Math. Pure Appl 23*, 37.

Lightman, A. P., W. H. Press, R. H. Price, and S. A. Teukolsky (1975). *Problem Book in Relativity and Gravitation*. Princeton University Press, Princeton.

Lindblom, L., M. A. Scheel, L. E. Kidder, R. Owen, and O. Rinne (2006). A new generalized harmonic evolution system. *Class. Quantum Grav. 23*, S447.

Linde, A. D. (1982). A new inflationary universe scenario: A possible solution of the horizon, flatness, homogeneity, isotropy and primordial monopole problems. *Phys. Lett. B 108*, 389.

Lousto, C. O. and J. Healy (2016). Unstable flip-flopping spinning binary black holes. *Phys. Rev. D 93*, 124074.

Lousto, C. O. and Y. Zlochower (2011). Hangup kicks: Still larger recoils by partial spin/orbit alignment of black-hole binaries. *Phys. Rev. Lett. 107*, 231102.

Lousto, C. O., Y. Zlochower, and M. Campanelli (2017). Modeling the black hole merger of QSO 3C 186. *Astrophys. J. 841*, L28.

Lovelace, G., R. Owen, H. P. Pfeiffer, and T. Chu (2008). Binary-black-hole initial data with nearly-extremal spins. *Phys. Rev. D 78*, 084017.

May, M. M. and R. H. White (1966). Hydrodynamic calculations of general-relativistic collapse. *Phys. Rev. 141*, 1232.

Merritt, D. and R. D. Ekers (2002). Tracing black hole mergers through radio lobe morphology. *Science 297*, 1310.

Misner, C. W., K. S. Thorne, and J. A. Wheeler (2017). *Gravitation* (2nd edn). Freeman, New York.

Montero, P. J. and I. Cordero-Carrión (2012). The BSSN equations in spherical coordinates without regularization: Vacuum and non-vacuum spherically symmetric systems. *Phys. Rev. D 85*, 124037.

Moore, T. A. (2013). *A General Relativity Workbook*. University Science Books, Mill Valley.

Mroué, A. H., M. A. Scheel, B. Szilágyi, H. P. Pfeiffer, M. Boyle, D. A. Hemberger, L. E. Kidder, G. Lovelace, S. Ossokine, N. W. Taylor, A. Zenginoğlu, L. T. Buchman, T. Chu, E. Foley, M. Giesler, R. Owen, and S. A. Teukolsky (2013). Catalog of 174 binary black hole simulations for gravitational wave astronomy. *Phys. Rev. Lett. 111*, 241104.

Nakamura, T., K. Oohara, and Y. Kojima (1987). General relativistic collapse to black holes and gravitational waves from black holes. *Prog. Theor. Phys. Suppl. 90*, 1.

Newman, E. T. and R. Penrose (1968). New conservation laws for zero rest-mass fields in asymptotically flat space-time. *Proceedings of the Royal Society of London Series A* **305**, 175.

Painlevé, P. (1921). Le mécanique classique et la théorie de la relativité. *L'Académie des sciences 173*, 677.

Pan, Y., A. Buonanno, A. Taracchini, L. E. Kidder, A. H. Mroué, H. P. Pfeiffer, M. A. Scheel, and B. Szilágyi (2014). Inspiral-merger-ringdown waveforms of spinning, precessing black-hole binaries in the effective-one-body formalism. *Phys. Rev. D 89*, 084006.

Paschalidis, V., Z. B. Etienne, R. Gold, and S. L. Shapiro (2013). An efficient spectral interpolation routine for the TwoPunctures code. *arXiv e-prints*, arXiv:1304.0457.

Petrich, L. I., S. L. Shapiro, and S. A. Teukolsky (1985). Oppenheimer–Snyder collapse with maximal time slicing and isotropic coordinates. *Phys. Rev. D 31*, 2459.

Poisson, E. (2004). *A Relativist's Toolkit: The Mathematics of Black-Hole Mechanics*. Cambridge University Press, Cambridge.

Press, W. H., S. A. Teukolsky, W. T. Vetterling, and B. P. Flannery (2007). *Numerical Recipes in C++: The Art of Scientific Computing* (3rd edn). Cambridge University Press, Cambridge.

Pretorius, F. (2005a). Evolution of binary black-hole spacetimes. *Phys. Rev. Lett. 95*, 121101.

Pretorius, F. (2005b). Numerical relativity using a generalized harmonic decomposition. *Class. Quantum Grav. 22*, 425.

Radice, D., A. Perego, F. Zappa, and S. Bernuzzi (2018). GW170817: Joint constraint on the neutron star equation of state from multimessenger observations. *Astrophys. J. Lett. 852*, L29.

Reinhart, B. (1973). Maximal foliations of extended Schwarzschild space. *J. Math. Phys. 14*, 719.

Rezzolla, L., E. R. Most, and L. R. Weih (2018). Using gravitational-wave observations and quasi-universal relations to constrain the maximum mass of neutron stars. *Astrophys. J. Lett. 852*, L25.

Rezzolla, L. and O. Zanotti (2013). *Relativistic Hydrodynamics*. Oxford University Press.

Ruchlin, I., J. Healy, C. O. Lousto, and Y. Zlochower (2017). Puncture initial data for black-hole binaries with high spins and high boosts. *Phys. Rev. D 95*, 024033.

Ruiz, M., S. L. Shapiro, and A. Tsokaros (2018). GW170817, general relativistic magnetohydrodynamic simulations, and the neutron star maximum mass. *Phys. Rev. D 97*, 021501.

Scheel, M. A., S. L. Shapiro, and S. A. Teukolsky (1995). Collapse to black holes in Brans–Dicke theory. I. Horizon boundary conditions for dynamical spacetimes. *Phys. Rev. D 51*, 4208.

Scheel, M. A., H. P. Pfeiffer, L. Lindblom, L. E. Kidder, O. Rinne, and S. A. Teukolsky (2006). Solving Einstein's equations with dual coordinate frames. *Phys. Rev. D 74*, 104006.

Schutz, B. F. (1980). *Geometrical Methods of Mathematical Physics*. Cambridge University Press, Cambridge.

Schutz, B. F. (2009). *A First Course in General Relativity* (2nd edn). Cambridge University Press, Cambridge.

Schwarzschild, K. (1916). Über das Gravitationsfeld eines Massenpunktes nach der Einsteinschen Theorie. *Sitzber. Deut. Akad. Wiss. Berlin, Kl. Math.-Phys. Tech.*, 189.

Seidel, E. and W.-M. Suen (1992). Towards a singularity-proof scheme in numerical relativity. *Phys. Rev. Lett. 69*, 1845.

Shapiro, S. L. and S. A. Teukolsky (1985a). Relativistic stellar dynamics on the computer. I. Motivation and numerical method. *Astrophys. J. 298*, 34.

Shapiro, S. L. and S. A. Teukolsky (1985b). Relativistic stellar dynamics on the computer. II. Physical applications. *Astrophys. J. 298*, 58.

Shapiro, S. L. and S. A. Teukolsky (1992). Collisions of relativistic clusters and the formation of black holes. *Phys. Rev. D 45*, 2739.

Shibata, M. (1999a). Fully general relativistic simulation of coalescing binary neutron stars: Preparatory tests. *Phys. Rev. D 60*, 104052.

Shibata, M. (1999b). Fully general relativistic simulation of merging binary clusters: Spatial gauge condition. *Prog. Theor. Phys. 101*, 1199.

Shibata, M. (2016). *Numerical Relativity*. World Scientific Publishing, Singapore.

Shibata, M. and T. Nakamura (1995). Evolution of three-dimensional gravitational waves: Harmonic slicing case. *Phys. Rev. D 52*, 5428.

Shibata, M., K. Uryū, and J. L. Friedman (2004). Deriving formulations for numerical computation of binary neutron stars in quasicircular orbits. *Phys. Rev. D 70*, 044044.

Shibata, M., S. Fujibayashi, K. Hotokezaka, K. Kiuchi, K. Kyutoku, Y. Sekiguchi, and M. Tanaka (2017). Modeling GW170817 based on numerical relativity and its implications. *Phys. Rev. D 96*, 123012.

Smarr, L. L. (1979). Gauge conditions, radiation formulae, and the two black hole collision. In L. L. Smarr (ed.), *Sources of Gravitational Radiation*, pp. 245. Cambridge University Press, Cambridge.

Smarr, L. and J. W. York, Jr. (1978). Radiation gauge in general relativity. *Phys. Rev. D 17*, 1945.

Starobinsky, A. A. (1980). A new type of isotropic cosmological models without singularity. *Phys. Lett. B 91*, 99.

Taracchini, A., A. Buonanno, Y. Pan, T. Hinderer, M. Boyle, D. A. Hemberger, L. E. Kidder, G. Lovelace, A. H. Mroué, H. P. Pfeiffer, M. A. Scheel, B. Szilágyi, N. W. Taylor, and A. Zenginoglu (2014). Effective-one-body model for black-hole binaries with generic mass ratios and spins. *Phys. Rev. D 89*, 061502.

Taylor, E. D. and J. A. Wheeler (2000). *Exploring Black Holes*. Addison Wesley Longman. San Francisco.

Taylor, J. H. and J. M. Weisberg (1982). A new test of general relativity – gravitational radiation and the binary pulsar PSR 1913+16. *Astrophys. J. 253*, 908.

Teukolsky, S. A. (2000). On the stability of the iterated Crank–Nicholson method in numerical relativity. *Phys. Rev. D 61*, 087501.

Thierfelder, M., S. Bernuzzi, and B. Brügmann (2011). Numerical relativity simulations of binary neutron stars. *Phys. Rev. D 84*, 044012.

Thornburg, J. (1987). Coordinates and boundary conditions for the general relativistic initial data problem. *Class. Quantum Grav. 4*, 1119.

Thorne, K. S. and D. MacDonald (1982). Electrodynamics in curved spacetime − 3+1 formulation. *Mon. Not. Roy. Astron. Soc. 198*, 339.

Toro, E. F. (1999). *Riemann Solvers and Numerical Methods for Fluid Dynamics: A Practical Introduction*. Springer-Verlag, Berlin.

van Meter, J. R., J. G. Baker, M. Koppitz, and D.-I. Choi (2006). How to move a black hole without excision: Gauge conditions for the numerical evolution of a moving puncture. *Phys. Rev. D 73*, 124011.

Wald, R. M. (1984). *General Relativity*. The University of Chicago Press, Chicago.

Weinberg, S. (1972). *Gravitation and Cosmology: Principles and Applications of the General Theory of Relativity*. Wiley, New York.

Wilson, J. R. and G. J. Mathews (1995). Instabilities in close neutron star binaries. *Phys. Rev. Lett. 75*, 4161.

York, Jr., J. W. (1971). Gravitational degrees of freedom and the initial-value problem. *Phys. Rev. Lett. 26*, 1656.

York, Jr., J. W. (1979). Kinematics and dynamics of general relativity. In L. L. Smarr (ed.), *Sources of Gravitational Radiation*, pp. 83. Cambridge University Press, Cambridge.

York, Jr., J. W. (1999). Conformal 'thin-sandwich' data for the initial-value problem of general relativity. *Phys. Rev. Lett. 82*, 1350.

Zangwill, A. (2013). *Modern Electrodynamics*. Cambridge University Press, Cambridge.

Zlochower, Y. and C. O. Lousto (2015). Modeling the remnant mass, spin, and recoil from unequal-mass, precessing black-hole binaries: The intermediate mass ratio regime. *Phys. Rev. D 92*, 024022.

Index